GENETIC
RESEARCH IN
PSYCHIATRY

GENETIC
RESEARCH
IN
PSYCHIATRY

Edited by Ronald R. Fieve, David Rosenthal, and Henry Brill

Proceedings of the Sixty-third Annual Meeting of the
American Psychopathological Association

The Johns Hopkins University Press
Baltimore and London

Copyright © 1975 by The Johns Hopkins University Press
All rights reserved. No part of this book may be reproduced
or transmitted in any form or by any means, electronic or
mechanical, including photocopying, recording, xerography, or
any information storage or retrieval system without permission
in writing from the publishers.
Manufactured in the United States of America

The Johns Hopkins University Press, Baltimore, Maryland 21218
The Johns Hopkins University Press Ltd., London

Library of Congress Catalog Card Number 74–24394
ISBN 0–8018–1660–2

Library of Congress Cataloging in Publication Data

American Psychopathological Association.
    Genetic research in psychiatry.

    Includes bibliographies and indexes.
    1. Mental illness—Genetic aspects—Congresses.
I. Fieve, Ronald R., ed. II. Rosenthal, David, ed.
III. Brill, Henry, 1906– ed. IV. Title.
[DNLM: 1. Mental disorders—Familial and genetic—
Congresses. W1 AM7299 63d 1973 / WM100 A512g 1973]
RC455.4.B5A45 1975   616.8'9'042   74–24394
ISBN 0–8018–1660–2

CONTENTS

Foreword ix
Preface xiii

PART I: The Biochemical Basis of Genetic Disorders
1. Special Award Lecture: Biochemical Genetics and Psychiatry
   Arno G. Motulsky and Gilbert S. Omenn 3
2. The Paul H. Hoch Award Lecture: Progress Toward
   an Understanding of the Biological Substrates of Schizophrenia
   Seymour S. Kety 15
3. Genetic Aspects of Catecholamine Synthesis
   Jack D. Barchas, Roland D. Ciaranello, Seymour Kessler,
   and David A. Hamburg 27

PART II: Genetic Studies of Criminality and Psychopathy
4. Extra Chromosomes and Criminality
   Seymour Kessler 65
5. Cytogenetic and Dermatoglyphic Studies in Sexual Offenders,
   Violent Criminals, and Aggressively Behaved Temporal
   Lobe Epileptics
   Lawrence Razavi 75
6. An Adoptive Study of Psychopathy: Preliminary Results
   from Arrest Records and Psychiatric Hospital Records
   Raymond R. Crowe 95
7. Registered Criminality in the Adoptive and Biological
   Parents of Registered Male Criminal Adoptees
   Barry Hutchings and Sarnoff A. Mednick 105
8. Discussion of Genetic Studies of Criminality and
   Psychopathy
   Lee N. Robins 117

PART III: Genetic Studies in Schizophrenia
9. Schizophrenic Spectrum Disorders in the Families of
   Schizophrenic Children
   Lauretta Bender 125
10. On the Possible Magnitudes of Selective Forces
   Maintaining Schizophrenia in the Population
   Kenneth K. Kidd 135

11. Mental Illness in the Biological and Adoptive Families of
    Adopted Individuals Who Have Become Schizophrenic:
    A Preliminary Report Based on Psychiatric Interviews
    *Seymour S. Kety, David Rosenthal, Paul H. Wender,
    Fini Schulsinger, and Bjørn Jacobsen*   147
12. Schizophrenia and the Schizoid: The Problem for
    Genetic Analysis
    James Shields, Leonard L. Heston, and Irving I. Gottesman   167
13. Discussion: The Concept of Subschizophrenic Disorders
    David Rosenthal   199
14. Discussion: Papers on the Genetics of Schizophrenia
    H. Warren Dunham   209

PART IV: Genetic Studies in Manic-Depressive Illness
15. Linkage Studies in Affective Disorders: The Xg Blood
    Group and Manic-Depressive Illness
    Julien Mendlewicz, Joseph L. Fleiss, and Ronald R. Fieve   219
16. The Familial History in Sixteen Males with
    Bipolar Manic-Depressive Illness
    H. Von Greiff, Paul R. McHugh, and Peter E. Stokes   233
17. A Dominant X-Linked Factor in Manic-Depressive Illness:
    Studies with Color Blindness
    *Ronald R. Fieve, Julien Mendlewicz, John D. Rainer,
    and Joseph L. Fleiss*   241

PART V: Genetic Studies in Other Disorders
18. The Transmission of Alcoholism
    Theodore Reich, George Winokur, and J. Mullaney   259
19. Genetic Studies of Hyperactive Children: Psychiatric
    Illness in Biologic and Adopting Parents
    Dennis P. Cantwell   273

PART VI: Social Implications of Genetic Theory
20. Presidential Address: Nature and Nurture as Political Issues
    Henry Brill   283
21. Genetic Knowledge and Heredity Counseling:
    New Responsibilities for Psychiatry
    John D. Rainer   289

Author Index   297

Subject Index   299

American Psychopathological Association

Officers for 1972–1973

Henry Brill, M.D., *President*	West Brentwood, N.Y.
Max Fink, M.D., *President-Elect*	New York, N.Y.
Charles Shagass, M.D., *Vice President*	Philadelphia, Pa.
Jonathan O. Cole, M.D., *Secretary*	Boston, Mass.
Donald F. Klein, M.D., *Treasurer*	Glen Oaks, N.Y.
Alfred M. Freedman, M.D., *Councillor*	New York, N.Y.
Seymour Fisher, M.D., *Councillor*	Boston, Mass.

Committee on Program

Ronald R. Fieve, M.D., *Chairman*	New York, N.Y.
John D. Rainer, M.D.	New York, N.Y.
George Winokur, M.D.	Iowa City, Iowa
David Rosenthal, M.D.	Bethesda, Md.

Arrangements

Robert L. Spitzer, M.D	New York, N.Y.

Publicity

David L. Dunner, M.D.	New York, N.Y.

FOREWORD

This noteworthy volume, marking the first full-scale symposium on genetics sponsored by the venerable American Psychopathological Association, comes at an important juncture in psychiatric research. Its contents look back to the early difficult days of genetics in psychiatry, difficult because of the limitations which then obtained in technique, in diagnostic sophistication, and in basic conceptualization. The pioneer investigators in psychiatric genetics did most of their work before the great advances in human chromosome research, in the biochemistry of genes and their products, and in mathematical and statistical methods. Diagnostically, by and large, they followed classical nosology and were often limited in their explorations to typical severe hospitalized patients. In their conceptual framework, they battled in the context of the nature-nurture controversy, having to thread their way between oversimplified theories on either side. The analysis of interaction between genetic and environmental influences based on methods for dividing population variance into its components was not fully developed or generally applied. What these early researchers did have at their disposal included the framework of Mendelian genetics with some attention to polygenes, techniques of ascertainment and of age correction, access to records, time, patience, and clinical acumen. Yet much of what they did foreshadowed the current surging of interest in psychiatric genetics, and without their work this book could not have been produced.

To be sure, workers in psychiatric genetics today are still faced with some of the same difficulties that confronted their forebears. Techniques of human chromosome identification have advanced to the point where each individual chromosome can be identified by its size, shape, and characteristic bands appearing under particular staining methods. Yet, as indicated by contemporary discussion in the field of schizophrenia and of criminality and psychopathy, the application of human chromosome knowledge to psychiatric states is both meager and controversial. Some persons have reported an excess of chromosomal aneuploidy among populations of schizophrenic patients or have described schizophrenia-like symptoms in individuals with extra chromosomes, particularly the X chromosome. Others have detected chromosomal mosaicism in populations of schizophrenic patients and possible clinical differences in those with chromosomal aneuploidy. With regard to criminality and psychopathy, there are many unresolved questions associated with chromosomal influence, particularly those relating to males with an extra Y chromosome. In male homosexuality there seems to be no typical chromosomal disorder, although certain male patients with chromosomal aneuploidy exhibit various forms of atypical sexual behavior. Since there is no reason to disbelieve a heterogeneous etiology, there has been speculation that the association of homosexuality with late maternal age indicates the role of a chromosomal disarrangement in some instances. And the intriguing hints regarding the effect of chromosomal aneuploidy both on intellectual function, as furnished by the observations on decreased space-form discrimination in girls with Turners syndrome, and on ego control, as evidenced by impulsive behavior in some XYY males, only indicate how fragmentary still is the application of the new chromosomal techniques to problems of human behavior in psychiatric disorder.

In the field of biochemical genetics, there is increasing motivation today to think and work toward a genetic neurochemistry. The relation of gene mutations to metabolic pathways goes as far back as Garrod, who described inborn errors of metabolism in the early 1900's. Subsequent theories related gene activity to enzyme production. New techniques of enzyme and other protein identification are now being exploited, and the variations which relate to the organism's ability to deal with the impact of environmental stress are particularly instructive.

Mathematical techniques today are more sophisticated than earlier techniques yielding only twin concordance rates or family risk figures, which many persons somehow treated as if they were baseball scores, indicating the relative strength of two teams, but nevertheless specific methods of fitting empirical facts to genetic models still require much reservation when applied to psychiatric disorders. Many assumptions and approximations have to be made, particularly about asymptomatic carriers, about environmental variation, and about significant environmental factors in the gene-environmental interaction. Progress in the biochemical area ought to make it more possible to evaluate the distribution of mental disorders of specific kinds in the population.

With the many advances in genetics, it has still not been possible to associate mental disorder with a specific gene behaving in a Mendelian fashion. Perhaps biochemical findings will establish such a pattern; meanwhile one of the closest approaches to this goal is represented by the studies of genetic linkage in manic-depressive psychosis which are currently reported. While the results of these analyses are not as strong as those in some other well-established linkage groups, these findings are probably the first time a human behavior disorder has been tentatively pinpointed to a specific locus in the genome. The data are certainly diluted because of diagnostic and perhaps etiological heterogeneity, but like all genetic studies they represent a first approximation to what may be the ultimate picture. It would be valuable to find ways of combining data on enzyme deficiencies in mental disorder with new cell hybridization techniques and thus provide another method for determining genetic linkage.

The diagnostic problems which earlier investigators may have glossed over in their pioneering studies have become important indeed, and their resolution is not yet in clear sight. However, one of the goals of psychiatric genetics actually is to develop methods for making clinically useful distinctions of a diagnostic nature which have also an etiological basis. Genetic studies have to start with a wide net, using subjects diagnosed on symptomatic, social, or other practical bases; subsequently, any heterogeneity may be established as the studies themselves come to distinguish one or more clearly defined groups in the process of subdivision and elimination. Another technique for the resolution of heterogeneity is represented by cross-cultural studies, noting variations in the frequency of a syndrome across different populations. If these variations are chiefly determined by cultural and environmental factors, there is less of a case for genetic homogeneity than if they are tied to population-genetic variables such as consanguinity, genetic drift, or selection. Studies of twins, families, and adoption can also attempt to distinguish between various types of schizophrenia, for example, or to spread the symptomatic types over a wide spectrum.

We still have not the laboratory techniques to answer fully the questions left us; we still have not all the data to feed into the equations which may help us to fit our theoretical models; we still need to cooperate more and more with clinicians who are devising better scales for rating diagnostic categories. Nevertheless, one gets the impression that both psychiatrists and geneticists have expanded their way of thinking. Genetics today admits regulation by the environment, from the molecular up to the social and evolutionary levels. Psychiatry, in turn, concerned with understanding human development, sees no loss but only a gain through including hypotheses and experiments relating to inborn dispositions. Standing on the shoulders of their predecessors, some of the investigators represented in this volume have suggested biochemical and cytological markers which may pave the way for identification of individuals predisposed to mental illness and may make it possible finally to resolve some of the current debates over diagnosis and classifications. It should then be possible to match the data to the most precise genetic analysis and to achieve a better biological synthesis for the etiology of the various syndromes under study. At the same time, longitudinal investigations of highly predisposed children (for example, offspring of affected parents) and the kind of adoption studies represented in the chapters that follow may also help to meet the objections of those who feel the environment is being neglected; indeed, they should help to refine our knowledge of environmental influences as they interact with genetic liability factors. Past, present, and future—they are all represented in this book, together with a

few warnings to temper enthusiasm with responsibility, to avoid political misuse, and to do no unwitting harm to patients and public by premature, sensationalistic, or anxiety-provoking statements. The science-
fiction of today is the science of tomorrow, but human and social values
have long evolutionary histories and ought to be guarded and preserved.

Here then is the present state of psychiatric genetics, a modern and
high-level presentation of the best in current thought—the present—drawing from the past and looking toward the future.

JOHN D. RAINER

PREFACE

It is fitting that the American Psychopathological Association should, for the first time, devote an entire symposium to the contributions of genetic research to the growing understanding of mental disorder. This is due to the relative paucity of papers in this area in the past and to the numerous advances in the areas of schizophrenia, mental deficiency, manic-depressive disease, and psychopathy during the past decade which have illuminated genetic mechanisms involved in these disorders. It is only recently that genetic research in psychopathology has gained widespread acceptance. At the turn of the century, before the rediscovery of Mendel, when neuropathologists were speculating about the inheritance of "damaged germ plasm" (e.g., Van Gieson 1898), genetic research in American psychiatry fell to a position of obscurity and disrepute in favor of the psychodynamic approach. While there were a few American researchers (e.g., Rosanoff 1912) who did apply Mendelian principles to the study of psychopathology in the early 1900's, this line of research did not gain popularity.

Within the American Psychopathological Association, Franz Kallmann was the lone contributor of genetics papers to the symposia until the middle 1960's. During the period from 1945 to 1966, Kallmann gave six papers on improvements in genetic methodology for studying psychiatric disorders; in 1947 he discussed the contingency method of estimating the incidence of a disease as it occurs in blood relatives of the proband compared to equivalent nonconsanguineous groups, and he also reported his work with the twin study method for establishing a genetic factor in psychiatric disorders. Kallmann later reported on genetic research in manic-depressive psychosis with family and twin studies in 1954 and continued to use these methodologies in his classic studies of schizophrenia, manic-depressive illness, and homosexuality. Kallmann's 1962 paper on the future of psychiatric genetics, in which he discussed molecular genetics in mongolism and sexual abnormalities, served with his other papers as a forerunner for later papers on cytogenetics by Lissy Jarvik in the 1969 volume and John Money in the 1970 volume. Thus, it can be said that Kallmann's early genetic studies, carried out when genetic research in psychopathology was looked down upon as a poor and disreputable cousin of psychodynamic theories, have been a principal force behind the renaissance of genetic research leading to the present symposium.

Psychopathology, as part of the science of man, presents unique challenges to the geneticist in the sense that he can observe in the human laboratory, but not manipulate. There can be no experimental manipulation of individual matings; the complete pedigree of most individuals is to some degree unknown or difficult to obtain; and in the nature of things there is even some uncertainty about the paternity of any given individual. Thus the "science of matings" is confronted with a situation in which matings can neither be produced at will nor observed with complete accuracy. These inherent limitations stimulated the development of the contingency and twin study methods which Kallmann helped to introduce in the United States as a first step toward the production of dependable genetic data.

The variety of new techniques and approaches represented in this volume can be viewed in the context of continuing and increasingly sophisticated attempts to grapple with these same limitations. The adoption techniques of studying both the biological and adoptive families of individuals adopted in infancy allow for a four-way comparison relating incidence of the disorder in the natural and adoptive parents to incidence in the offspring. This provides a tool for separating the effects of genetic endowment from environmental influences on the development of an individual which is more powerful than earlier techniques, such as family and twin studies. Variations of the adoption techniques are used in this volume by Kety and his group in their studies of schizophrenia, by Crowe and by Hutchings in their papers on criminal behavior, and by Cantwell in his paper on hyperactive children. Linkage studies have added a powerful tool to the determination of the mode of inheritance. Fieve and Mendlewicz in their papers attempt to establish a linkage of the locus of manic-depressive illness to the loci of known genetic markers by studying the co-occurrence of manic-depression with the known trait in family pedigrees, while Greiff contests the hypothesis of X-linked dominant inheritance.

Genetic studies have added impetus to the claim that there is a spectrum of schizophrenic disorders of common etiology, just as a variety of clinical disorders have been associated with the defective enzyme of the Lesch–Nyhan syndrome. Kety and his group attempt to define the schizophrenic spectrum on the basis of considerable data obtained through the adoptee technique. Shields also deals with the spectrum problem in a thorough review, as does Bender in an unusual longitudinal study.

Psychopathologists have been plagued from time immemorial by the nature-nurture issue. It may be that genetic advances over the past decade will have helped in great part to resolve this debate by offering positive evidence for a genetic component in specific disorders, and by helping to define the boundaries of mental disorders more accurately. In no small measure, these advances have been made possible by the development of powerful analytic methods such as those described and used in the papers by Fieve, Kidd, Mendlewicz, and Reich.

Studies of schizophrenia, alcoholism, psychopathy, and manic-depressive illness as illustrated in this volume now lead one to conclude that there is indeed a demonstrable genetic factor present in all of these illnesses. Although this claim does not surprise most biological psychiatrists who have researched these illnesses and treated them with this hypothesis in mind for over half a century, the fact remains that only recently have we been able to obtain solid empirical evidence. The real consequence then of pinning down a genetic or biological component in these illnesses is to help separate out homogeneous populations of psychiatric disorders that can in turn be studied more definitively for specific modes of transmission of the illness and for biochemical abnormalities in different body components. Finally, the further behavioral-biochemical study of the individual now comes much more within the grasp of the clinical researcher when he has the firm knowledge that he is dealing with subtypes of illnesses alike not only in phenotype, i.e., the psychopathology that is part of the present and past history of the individual, but also alike in terms of possible genotype, when a specific mode of transmission of a trait is defined in a given pedigree with the aid of a genetic marker.

Researchers in clinical psychiatry have labored now for decades testing various body fluids for abnormal components, only to find more often than not that the differences detected could be ascribed to artifacts due to drug therapy, diet, activity, but most important, lack of homogeneity in diagnosis. The new genetic techniques illustrated in this volume now enable the clinical researcher to establish those homogeneous subpopulations which can be intensively researched from a variety of approaches, including specific studies on genetic transmission of subgroups of disorders and specific biochemical hypothesis-testing done on subgroups that are defined by genetic familial studies.

In closing, we would like to express our gratitude to David A. Zubin for coordinating and carrying out much of the editorial work and to Judith Roth for her help in arranging the program and publicizing the meeting.

R. F., D. R., and H. B.

References

Rosanoff, A. J. On the inheritance of the neuropathic condition. *State Hospitals Bulletin.* 1912, n.s. **5**, 278–292.

Van Gieson, I. Correlation of sciences in the investigation of nervous and mental diseases. *Archives of Neuropathology and Psychopathology*, 1898, **1**, 25–262.

I

THE BIOCHEMICAL BASIS OF GENETIC DISORDERS

1 SPECIAL AWARD LECTURE: BIOCHEMICAL GENETICS AND PSYCHIATRY

Arno G. Motulsky and
Gilbert S. Omenn

Family studies and twin studies point to a prominent role of genetic factors in many psychiatric disorders in man (Rosenthal 1970; Slater & Cowie 1971; Erlenmeyer-Kimling & Jarvik 1963; Shields 1962). Recent investigations of the biological and adoptive families of schizophrenics and of others with psychopathology (Kety, Rosenthal, Wender, & Schulsinger 1972) have finally served to ferret out much of the common intrafamilial environment from transmitted inherited factors. Thus, the emphasis can shift now from an extreme nature vs. nurture controversy to identification of the biochemical basis of genetic predispositions to psychiatric disorders. Most important in such investigations will be recognition of genetic heterogeneity—that is, multiple different mechanisms that lead to similar clinical phenotypes. Etiologic heterogeneity is well-known for the many specifically-diagnosable causes of anemia and of mental retardation, as examples.

With genetic counseling for psychiatric disorders, there is hope for earlier detection of those at high risk; psychotherapy or drug therapy might be more effective if initiated as early as possible. While emphasis on genetic-environmental interaction appears warranted, rational approaches are hampered by the limited knowledge of underlying biochemical abnormalities and of the presumed heterogeneity among those whose family histories suggest they are genetically predisposed. Nevertheless, knowledge of basic biochemistry of gene action, clues from certain inherited diseases, and analyses of the actions of psychopharmacologic agents offer promise of bridging the chasm between brain biochemistry and observed behaviors.

Basic Biochemical Genetics of the Human Brain

The molecular picture of gene action drawn from studies of brain and other tissues of many organisms is highly generalized. Genetic information is

Arno G. Motulsky, M.D., and Gilbert S. Omenn, M.D., Division of Medical Genetics, Departments of Medicine and Genetics, University of Washington, Seattle, Washington 98195.
Supported by grant GM 15253 and a Research Career Development Award (G.S.O.) from the U.S. Public Health Service.

coded in the DNA of the chromosomes which are passed from generation to generation in egg and sperm cells. Genetic information in somatic cells is expressed via RNA messengers as protein products. Many regulatory steps affect the synthesis and degradation of these protein products. In fact, even though all cells, including neurons and neuroglial cells, contain the same complement of DNA, the regulatory processes of differentiation lead to different patterns of gene expression in different tissues. The brain constitutes 2 to 3 percent of body weight, yet its metabolism consumes up to 50 percent of the resting energy and oxygen supply. The "resting" state of neurons is characterized by intense rhythmic and spontaneous electrophysiological activity and by active processes of biosynthesis and transport of proteins. Biochemical techniques are now available to examine both the RNA and protein products of genes in the brain.

Transcription of DNA   How much of the genetic information coded in the DNA is expressed in a given tissue? The technique used in studies to answer this question involves preparation of radioactively labeled fragments of DNA from any tissue of the species being studied, followed by "hybridization" of the DNA with an excess of RNA from the tissue under investigation. If the DNA has been transcribed into complementary RNA messenger, this RNA messenger will form DNA/RNA hybrids with the labeled DNA; if the DNA was not expressed as RNA messenger in that tissue, there will be no RNA counterpart to form a hybrid and the radioactive DNA will be separable from that which forms hybrids. Such techniques have shown that about 10 percent of the genetic information of mouse DNA is expressed in tissues like liver, kidney, and spleen; twice as much of the DNA is utilized in the brain (Hahn & Laird 1971; Grouse, Chilton, & McCarthy 1972). In recent sudies of this type with human tissues, a similar proportion of the genetic information is used in liver, kidney, and spleen, but even higher proportions (as much as 45 percent) of the genetic information is expressed in adult brain cortex (Grouse, Omenn, & McCarthy 1973). The adult gives higher values than fetal brain, and cortex gives higher values than brain stem. These preliminary findings are highly provocative: the much greater diversity of RNA molecules in the brain cortex might reflect complex cognitive functions, including memory storage; certainly the development of the brain cortex is the most impressive anatomical feature of human brain evolution, and these biochemical findings suggest that such evolution of the cortex was accompanied by ever greater selection of human genes for brain functions. Presumably, most of these genes are inactive in cells of other tissues.

Protein Variation   The proteins of cells and of body fluids have been studied extensively to determine whether individuals differ in the properties of certain plasma proteins, enzymes, hemoglobins, blood groups, and other antigens. In fact, many proteins are found to have different properties in different individuals; for example, there are over 100 slightly different hemoglobin proteins. Some cause disease; most are innocuous. In the ABO blood groups and in several red cell enzyme and serum proteins, variants are very common (Giblett 1969). If more than one percent of a tested population has a particular variant protein, that protein is said to be polymorphic. Such proteins are very useful as genetic markers for population studies, paternity testing, and various analyses of gene action. Tests of a series of approximately twenty-five existing gene markers give a profile of individuals sufficiently variable that the probability of two nonidentical twins having the same genetic profile is about one in three billion (Omenn & Motulsky 1972b, Table 6), corresponding to the world's population. Thus, from even quite limited biochemical genetic analysis, we can corroborate the poet's intuition that each human is truly unique.

We have recently begun to apply to brain tissue the electrophoretic techniques which have uncovered polymorphic variation in about one-third of blood enzymes tested (Harris & Hopkinson 1972). If variants of key brain enzymes involved in intermediary metabolism and in neurotransmitter synthesis and breakdown could be found, characterization of the functional effects of such variants might provide a physiological link between the variation at the level of physical chemical properties and correlated behavioral differences. To the surprise of many geneticists, however, the initial screening studies have found none of the anticipated variation in key brain enzymes, like those involved in metabolism of glucose. No variation in this pathway was found in red cells, either. Instead of the expected one-third of enzymes being polymorphic, none of the thirteen genetic loci coding for the enzymes of the glycolytic pathway were variable at all. Presumably, there is strong selection against variants in this pathway. Variants presumably arose by mutation, but unlike in other pathways of metabolism did not survive. We have concluded (Cohen, Omenn, Motulsky, Chen, & Giblett 1973) that such "housekeeping" functions as glucose utilization for energy production are fixed in an optimal structural-functional relationship and are not the site of what the psychologists call the "plasticity" of the brain nor that of other tissues. As proteins involved in more specialized brain functions, including neurotransmitter-related enzymes, are investigated, there may be a greater chance of finding variation, variation significantly related to the individual differences we perceive in behavior.

Pharmacogenetic Investigations of Specific Enzymes   The electrophoretic methods in studies of enzyme variation have two important limitations. First, they miss altogether changes in proteins which do not introduce a change in net charge of the protein. Second, there is no simple way to relate the electrophoretic difference between two forms of the enzyme to differences in function of the enzyme in the brain *in vivo*. For example, we have discovered a highly polymorphic enzyme system in brain as well as other tissues, the mitochondrial malic enzyme. This enzyme interconnects the Krebs' cycle of glucose metabolism with lipid biosynthesis. Multiple biochemical analyses of the properties of the variant forms of the enzyme failed to demonstrate any significant functional differences in the test tube (Cohen & Omenn 1972). However, slight differences in enzyme activity within the normal range are the rule for enzyme variants and may explain much of normal variation for such enzymes (Motulsky 1970).

The pharmacogenetic approach draws upon common observations in clinical psychiatry and offers promise of overcoming both of the limitations of the electrophoretic method. The marked differences across individuals in their responses to psychoactive agents suggest strongly that there must be differences in the susceptibility of the brain to the action of the drug or differences in the way the drug is metabolized (Omenn & Motulsky 1973b). Although the laboratory methods are very tedious, involving numerous enzyme assays on a sample from each individual, significant variation in the susceptibility, for example, of the enzyme monoamine oxidase (MAO) to clinically utilized MAO inhibitors, would allow a clinical correlation. It may even be possible to test the effects of the MAO inhibitor on the MAO of blood platelets, rather than requiring brain tissue, but the identity of platelet and brain MAO remains to be established. In principle, each of the important enzymes of biosynthesis and degradation of such neurotransmitters as norepinephrine, dopamine, serotonin, acetylcholine, and gamma-aminobutyric acid may be studied in this fashion. There are drugs which specifically inhibit these enzymes, including some which have been used clinically (MAO inhibitors, alpha-methyldopa, dopa decarboxylase inhibitors).

Future Areas   Recent histofluorescent techniques allow the identification of neurotransmitters within particular nerve tracts (Hillarp, Fuxe, & Dahlstrom 1966). The chemicals thought to be the major neurotransmitters (norepinephrine, dopamine, serotonin, acetyl choline, GABA) probably account for less than one-fourth of the demonstrable synapses between neurons, leaving the likelihood that multiple other neurotransmitters may function in the brain. New methods of biochemical genetic analysis of cells of nervous system origin grown *in vitro* in tissue culture media may be very helpful in sorting out the individualized functions of different types of cells. The formation of synapses themselves requires some remarkable process of cell-cell recognition, presumably mediated by a mosaic of complex macromolecules, such as glycoproteins, in the cell surfaces. Techniques for isolating and analyzing these complex proteins are just becoming practical. Still another physiologically important area to be explored involves biological rhythms, some of which are mediated by circulating hormones.

Clues from Clinical Genetics

Considerable understanding of normal functions of the red blood cell, liver, kidney, and intestine has been deduced from analysis of the effects of inborn errors of metabolism (Stanbury, Frederickson, & Wyngaarden 1972). Such "experiments of Nature" have become useful in designing suitable diagnostic tests and, in some cases, rational therapies. Perhaps the most famous example is that of phenylketonuria (PKU). In this disease, severe mental retardation results from the damaging effects on the developing brain from toxic circulating concentrations of phenylalanine and its ketone metabolites. Phenylalanine is an essential amino acid needed for growth; it is an important constituent of proteins, and it is converted into tyrosine and other essential molecules. The disease PKU is due to a block in the step of conversion of phenylalanine to tyrosine; the phenylalanine builds up and damages key functions in the brain. There is nothing intrinsically the matter with the brain; the enzyme deficiency is in the liver, and the poisonous effect of the phenylalanine is analogous to the effects of lead poisoning on the brain. From such biochemical understanding, suitable diagnostic and therapeutic approaches could be designed. The measurement of blood phenylalanine concentration became a standard test for newborns, though not every child with high values actually has PKU. Most important, administration of a diet limited in phenylalanine can prevent the development of the mental retardation. The diet must provide enough phenylalanine to support growth, but not enough to accumulate and become toxic to the brain. It will be important to examine adult individuals who were helped by the PKU diet in childhood to see whether more subtle mental abnormality becomes manifest as psychoses or severe neuroses.

PKU is one of several dozen inborn errors of metabolism which result in mental retardation; some directly involve the brain, while many others appear to act in this indirect, toxic fashion. Three especially interesting examples of inborn errors intrinsic to the brain are Lesch–Nyhan syndrome, homocystinuria, and metachromatic leukodystrophy.

The Lesch–Nyhan syndrome consists of involuntary choreoathetoid movements, a bizarre self-destructive behavior with biting of the lips and fingertips, and severe gouty arthritis. Gout results from very high levels of uric acid, which, in turn, are due to specific deficiency of the enzyme HGPRT (Kelley & Wyngaarden 1972). This enzyme was long thought to be part of an insignificant side pathway in the metabolism of purines needed to make DNA and RNA. It was only with this "experiment of nature" that the consequences of deficiency of this enzyme activity were learned. The highest levels of this enzyme in the body are found normally in the basal ganglia of the brain, so that deficiency is consistent with the involuntary movement disorder. What remains to be understood is the most intriguing part of this disease: the relationship of the metabolic pathway or of uric acid levels to the violent behavior the patient directs at himself. The HGPRT enzyme is coded for by a gene on the X chromosome, so that only boys are affected, though females may be carriers.

Homocystinuria is an autosomal recessive disorder affecting the bones, blood vessels, eye, skin, and sometimes mental development. Affected men and women are tall and gangly, like the Marfan's syndrome, and are uniquely vulnerable to clotting in both veins and arteries. A particular enzyme is deficient, cystathionine synthetase, which produces the complex amino acid cystathionine. Although its functions are not known, cystathionine is found in the highest concentrations in the brain and may even serve as a neurotransmitter. Half of these patients are described as moderately mentally retarded. The important question is why only half? Would those who have normal IQ's have had superior IQ's if not for this genetic disorder? It is possible to answer that question by measuring the IQ's of the patients and siblings of individuals with homocystinuria to estimate the expected IQ for the patient. Homocystinuria is of interest for another reason. In several early reports an apparent excess of cases of schizophrenia was noted among the relatives of the patients. The relatives were probably carriers of the gene for homocystinuria, and it has been speculated that abnormal metabolism of methylated metabolites, such as occurs in homocystinuria, might occur in schizophrenia. Although the frequency of homocystinuria is on the order of one case in 40,000 people, even this low frequency corresponds to a prevalence of carriers of one per 100 population. In fact, for all of the autosomal recessive inborn errors of metabolism that can lead to mental retardation, and especially for those whose action is intrinsic to the brain, it is conceivable that individuals who are carriers and have these abnormal genes may be predisposed to minor psychological and psychiatric difficulties. Given the dozens of such inborn errors, each of which has a carrier frequency approaching one percent, it is likely that many of us are carriers for two or more potentially deleterious genes.

In addition to carriers, there may be individuals who have a different mutation of a gene causing less than complete deficiency of the enzyme. For example, complete deficiency of the enzyme cerebroside sulfatase results in accumulation of sulfatides in the brain and other tissues in infancy. Neurological and mental function disintegrates and leads to early death. However, an adult form of this disease, metachromatic leukodystrophy, has been reported in at least two dozen cases (Austin, Armstrong, Fouch, Mitchell, Stumpf, Shearer, & Briner 1968). In every case the adolescent or young adult was first diagnosed as having schizophrenia and was institutionalized with that diagnosis. Only later when neurological signs developed and when the brain was found to be grossly abnormal from accumulation of sulfatides was this specific diagnosis made.

This disease also serves as an example of what may be very many different specific entities among the causes of schizophrenia.

Amniocentesis and Prenatal Diagnosis   One of the most important developments in medical genetics has been the specific diagnosis of certain metabolic or chromosomal disorders early enough in pregnancy so that precise advice can be given about whether that particular fetus is affected or not (Nadler 1968; Omenn & Motulsky 1973a). In a chromosomal disorder, such as Down's syndrome or mongolism, where there is an extra chromosome in all cells, it is a straight-forward procedure to obtain fetal cells from the amniotic fluid and examine the chromosomes in a laboratory. For metabolic disorders, it is essential to be certain that the responsible enzyme can normally be detected in the amniotic fluid cells. Homocystinuria, metachromatic leukodystrophy, and the Lesch–Nyhan syndrome all can be diagnosed in this fashion. Another important example is Tay–Sachs disease, because it is found in such a disproportionately high frequency among Jews who are now being tested in several metropolitan areas. On the other hand, it is not feasible to test for PKU, because that enzyme is normally found in the liver and not in such cells as can be obtained from the amniotic fluid. Similarly, tests for sickle cell anemia require blood cells, while tests for another very common disease, cystic fibrosis, are impossible for lack of knowledge about any specific enzyme abnormality.

A number of interesting behavior syndromes have been associated with chromosomal disorders, including poor space-form perception in Turner's syndrome (45, XO), significant psychopathology in Klinefelter's syndrome (47, XXY), and slight but seemingly significant predisposition to criminal behavior in the XYY syndrome. Since each chromosome contains enough DNA for thousands of genes, we have no basis at present to correlate gross chromosomal abnormalities with any metabolic mechanisms for behavior abnormalities.

Biochemical Genetic Approaches to Major Psychiatric Disorders

Affective Disorders   According to the biogenic amine hypothesis (Schildkraut 1969), depression may be associated with decreased action of turnover of norepinephrine (NE) and serotonin, while manic states are associated with increased biogenic amine turnover. Pharmacologic agents which deplete NE from nerve terminals (reserpine) or interfere with its synthesis (alpha-methyl tyrosine, alpha-methyl dopa) may precipitate depression. Drugs which enhance biosynthesis of NE (L-Dopa) may induce hypomanic states, and agents which prolong the action of NE by inhibiting intraneuronal monoamine oxidase (MAO inhibitors) or the neuronal reuptake of NE released into the synapse (tricyclics) are effective antidepressants. Electro-convulsive shock also acts to increase tyrosine hydroxylase activity and NE turnover (Mussacchio, Julou, Kety, & Glowinski 1969). There may be genetically determined individual differences in any of these steps in the synthesis or turnover of this transmitter substance that could be revealed by analysis of the effects of pharmacologic agents. For example, groups of patients have been reported to be differentiated by their response to either MAO inhibitors or tricyclic compounds (Pare, Rees, & Sainsbury 1962; Angst 1964; Pare & Mack 1971). According to these authors, patients who respond to one class of antidepressant tended not to respond to the other. Patients showed the same pattern of responsiveness during a subsequent episode of depression which might have been precipitated by quite different life stresses. Relatives who had affective disorders shared the pattern of pharmacologic responsiveness or unresponsiveness. About 10 percent of patients with high blood pressure treated with reserpine develop depression (Harris 1957). The reserpine appears to unmask a predisposition to depression. ACTH and alpha-methyl dopa also precipitate depression in small percentages of patients given those drugs. By studying the biochemical sites of action of these drugs and by investigating families of individuals predisposed to such triggering of affective disorders we may learn more about the heterogeneous causes of affective disorders. Although the relationship of peripheral enzymes to brain enzymes remains to be established, preliminary experiments have already been reported for the catechol-O-methyl transferase of red blood cells (Cohn, Dunner, & Axelrod 1970; Dunner, Cohn, Gershon, & Goodwin 1971) and the monoamine oxidase of platelets (Murphy & Weiss 1972) in patients with manic-depressive illness.

Schizophrenia   At present we have only a few scattered leads for new investigations into schizophrenia. The examples of adult onset metachromatic leukodystrophy, acute intermittent porphyria, and Huntington's disease illustrates the need to ferret out heterogeneity in designing studies. Thus, intensive studies of a large kindred with several cases of schizophrenia might yield much more interpretable information than study of many more unrelated patients who may have many different underlying mechanisms. There are two threads of evidence to direct attention to the role of dopamine in schizophrenic patients. The active groups of potent antipsychotic phenothiazines, when viewed in a three-dimensional molecular model, appear to resemble the molecular conformation of dopamine (Horn & Snyder 1971). Also, the equivalence of D- and L-amphetamine in inducing schizophrenic-like psychosis points to involvement of dopamine (Angrist, Shopsin, & Gerhson 1971; Snyder, Taylor, Coyle, & Meyerhoff 1970). Given the evidence for genetic predisposition to schizophrenia (Rosenthal 1970, pp. 92–200), one may wonder whether individuals who have a schizophrenic-like reaction to amphetamines or LSD were genetically predisposed to such psychotic reactions and would have been at a relatively high risk for development of "spontaneous" schizophrenia.

Seizure Disorders   The normal pattern of electrical activity in the brain, as measured by the electroencephalogram (EEG), is determined almost entirely by multiple genetic factors. Several single-gene-mediated variants of the normal EEG have been described, affecting altogether about 15 percent of the general population (Vogel 1970). The clinical significance of these variants is unknown. Studies are needed to determine whether individuals with different baseline EEG patterns have different responses to various psychopharmacologic agents. In addition, there is urgent need to evaluate the current controversy that impulsive and violent behavior in some individuals may be due to nonmotor seizure disorders.

Minimal Brain Dysfunction—Hyperkinetic Children   This diagnostic label has been applied to as many as 5 to 10 percent of school children, especially boys, who have motor hyperactivity, distractibility, impulsivity, and a learning performance below objective expectations. The most important clue for investigation of underlying mechanisms is the evidence that some of the children, but clearly not the majority, respond strikingly to stimulant drugs such as amphetamines and methylphenidate. The variation in responsiveness to these drugs may reflect differences in the metabolism of the drug, in susceptibility to the drug action at the level of some cell enzyme or cell membrane, or in the underlying processes which lead to the behavioral problems (Omenn 1973).

Alcoholism   The most commonly abused addicting drug, of course, is ethanol. Individuals differ markedly in tolerances for alcohol and in susceptibility to its effects. Recent studies with half siblings have shown that the excess familial incidence of alcoholism is among the biologically related and not the adoptive family members, pointing to genetic rather than environmental influences (Schuckit 1972). Genetic factors could influence variation in the acute intoxicating effects of alcohol, in the likelihood of chronic addiction, and in the risk of medical and behavioral complications (Omenn & Motulsky 1972a). Genetic factors are almost entirely responsible for individual differences in the rate of elimination of ethanol from the blood (Vessell, Page, & Passananti 1971). Eskimos and Indians metabolize ethanol less rapidly than do Caucasians in Western Canada (Fenna, Mix, Schaefer, & Gilbert 1971). Most Oriental adults and infants show facial flushing and increased pulse pressure after doses of alcohol which have little or no effect on Caucasians (Wolff 1972). Among children born to mothers who are chronic alcoholics, a striking new syndrome has been recognized (Jones, Smith, Ulleland, & Streissguth 1973). These children have retarded intrauterine and postnatal physical and mental development, microcephaly, decreased width to the palpebral fissure causing the eyes to appear rounded, and variable limb and cardiac malfunctions. The toxic mechanism and genetic variation in susceptibility have yet to be investigated. These several observations related to social and medical consequences of alcoholism should stimulate more intensive biochemical and genetic studies into the crucial relationship between acute intoxicating effects and chronic addiction.

References

Angrist, B. M., Shopsin, B., & Gershon, S. The comparative psychotomimetic effects of stereoisomers of amphetamine. *Nature*, 1971, **234**, 152–153.

Angst, J. Antidepressiver Effekt und genetische Factoren. *Arzneimittelforschung. Supplement*, 1964, **14**, 496–500.

Austin, J., Armstrong, D., Fouch, S., Mitchell, C., Stumpf, D., Shearer, L., & Briner, O. Metachromatic leukodystrophy (MLD) VIII. MLD in adults; diagnosis and pathogenesis. *Archives of Neurology*, 1968, **18**, 225–240.

Cohen, B. T. W., & Omenn, G. S. Human malic enzyme: high frequency polymorphism of the mitochondrial form. *Biochemical Genetics*, 1972, 7, 303–311.

Cohen, B. T. W., Omenn, G. S., Motulsky, A. G., Chen, S.-H., & Giblett, E. R. Restricted variation in the glycolytic enzymes of human brain and erythrocytes. *Nature. New Biology*, 1973, **241**, 229–233.

Cohn, C. K., Dunner, D. L., & Axelrod, J. Reduced catechol-O-methyltransferase activity in red blood cells of women with primary affective disorder. *Science*, 1970, **170**, 1323–1324.

Dunner, D. L., Cohn, C. K., Gershon, E. S., & Goodwin, E. K. Differential catechol-O-methyltransferase activity in unipolar and bipolar affective illness. *Archives of General Psychiatry*, 1971, **25**, 348–353.

Erlenmeyer-Kimling, L., & Jarvik, L. F. Genetics and intelligence: a review. *Science*, 1963, **142**, 1477–1478.

Fenna, D., Mix, L., Schaefer, O., & Gilbert, J. A. L. Ethanol metabolism in various racial groups. *Canadian Medical Association Journal*, 1971, **105**, 472–475.

Giblett, E. R. *Genetic markers in human blood*. Philadelphia: F. A. Davis, 1969.

Grouse, L., Chilton, M-D., & McCarthy, B. J. Hybridization of ribonucleic acid with unique sequences of mouse deoxyribonucleic acid. *Biochemistry*, 1972, **11**, 798–805.

Grouse, L., Omenn, G. S., & McCarthy, B. J. Studies by DNA/RNA hybridization of the transcriptional diversity of human brain. *Journal of Neurochemistry*, 1973, **19**, 1063–1073.

Hahn, W. E. & Laird, C. D. Transcription of nonrepeated DNA in mouse brain. *Science*, 1971, **173**, 158–161.

Harris, T. H. Depression induced by Rauwolfia compounds. *American Journal of Psychiatry*, 1957, **113**, 950.

Harris, H., & Hopkinson, D. A. Average heterozygosity per locus in man: an estimate based on incidence of enzyme polymorphisms. *Annals of Human Genetics*, 1972, **36**, 9–20.

Hillarp, N. A., Fuxe, K., & Dahlstrom, A. Demonstration and mapping of central neurons containing dopamine, noradrenaline and 5-hydroxytryptamine and their reactions to psychopharmaca. *Pharmacological Review*, 1966, **18**, 727–741.

Horn, A. S., & Snyder, S. H. Chlorpromazine and dopamine: conformational similarities that correlate with the anti-schizophrenic activity of phenothiazine drugs. *Proceedings of the National Academy of Sciences (Washington, D.C.)*, 1971, **68**, 2325–2328.

Jones, K. L., Smith, D. W., Ulleland, C. N. & Streissguth, A. P. Patterns of malformation in offspring of chronic alcoholic mothers. *Lancet*, 1973, **1**, 1267–1271.

Kelley, W. N., & Wyngaarden, J. B. The Lesch–Nyhan Syndrome. In J. B. Stanbury, D. S. Frederickson, & J. B. Wyngaarden (Eds.), *The metabolic basis of inherited disease.* (3rd ed.) New York: McGraw–Hill, 1972, pp. 969–991.

Kety, S. S., Rosenthal, D., Wender, P. H., & Schulsinger, F. Mental illness in the biological and adoptive families of adopted schizophrenics. *American Journal of Psychiatry*, 1971, **128**, 82–91.

Motulsky, A. G. General remarks on genetic factors in anesthesia. *Humangenetik*, 1970, **9**, 246–249.

Murphy, D. L. & Weiss, R. Reduced monoamine oxidase activity in blood platelets from bipolar depressed patients. *American Journal of Psychiatry*, 1972, **128**, 1351–1357.

Mussachio, J. M., Julou, L., Kety, S. S., & Glowinski, S. S. Increase in rat brain tyrosine hydroxylase activity produced by electroconvulsive shock. *Proceedings of the National Academy of Sciences (Washington, D.C.)*, 1969, **63**, 1117–1119.

Nadler, H. L. Prenatal genetic diagnosis. *Pediatrics*, 1968, **42**, 912–918.

Omenn, G. S. Genetic issues in the syndrome of minimal brain dysfunction. *Seminars in Psychiatry*, 1973, **5**, 5–17.

Omenn, G. S., & Motulsky, A. G. Biochemical genetic approach to alcoholism. *Annals of the New York Academy of Sciences*, 1972, **197**, 16–23. (a)

Omenn, G. S., & Motulsky, A. G. Biochemical genetics and the evolution of human behavior. In L. Ehrman, G. S. Omenn, & E. Caspari (Eds.), *Genetics, environment and behavior: implications for educational policy.* New York: Academic Press, 1972, pp. 129–171. (b)

Omenn, G. S., & Motulsky, A. G. Intra-uterine diagnosis and genetic counseling: implications for psychiatry in the future. In D. A. Hamburg & H. K. H. Brodie (Eds.), *American handbook of psychiatry.* (3rd ed.) New York: Basic Books, 1973, in press. (a)

Omenn, G. S., & Motulsky, A. G. Pharmacogenetics (Psychopharmacogenetics). Yearbook of Drug Therapy, 1973, 5–26. (b)

Pare, C. M. B., & Mack, J. W. Differentiation of two genetically specific types of depression by the response to antidepressant drug. *Journal of Medical Genetics*, 1971, **8**, 306–309.

Pare, C. M. B., Rees, L., & Sainsbury, M. J. Differentiation of two genetically specific types of depression by response to anti-depressants. *Lancet*, 1962, **2**, 1340–1342.

Rosenthal, D. *Genetic theory and abnormal behavior.* New York: McGraw–Hill, 1970.

Schildkraut, J. J. Neuropsychopharmacology and the affective disorders. *New England Journal of Medicine*, 1969, **281**, 197–201, 248–255, 302–308.

Schuckit, M. A. Family history and half-sibling research in alcoholism. *Annals of the New York Academy of Sciences*, 1972, **197**, 121–124.

Shields, J. *Monozygotic twins brought up apart and brought up together.* London: Oxford University Press, 1962.

Slater, E., & Cowie, V. *The genetics of mental disorders*. New York: Oxford University Press, 1971.

Snyder, S. H., Taylor, K. M., Coyle, J. T., & Meyerhoff, J. L. The role of brain dopamine in behavioral regulation and the actions of psychotropic drugs. *American Journal of Psychiatry*, 1970, **127**, 199–207.

Stanbury, J. B., Frederickson, D. S., & Wyngaarden, J. B. *The metabolic basis of inherited disease*. (3rd ed.) New York: McGraw–Hill, 1972.

Vesell, E. S., Page, J. G., & Passananti, G. T. Genetic and environmental factors affecting ethanol metabolism in man. *Clinical Pharmacology and Therapeutics*, 1971, **12**, 192–201.

Vogel, F. The genetic basis of the normal human electroencephalogram (EEG). *Humangenetik*, 1970, **10**, 91–114.

Wolff, P. H. Ethnic differences in alcohol sensitivity. *Science*, 1972, **175**, 449–450.

Seymour S. Kety
Paul H. Hoch Award Lecturer, 1973

2 THE PAUL H. HOCH AWARD LECTURE: PROGRESS TOWARD AN UNDERSTANDING OF THE BIOLOGICAL SUBSTRATES OF SCHIZOPHRENIA

Seymour S. Kety

Paul Hoch was a perceptive and productive scholar of schizophrenia, contributing to our knowledge of its clinical form, recognizing the importance of psychopharmacology to its treatment and understanding, cognizant of both its biological substrates and experiential superstructure, concerned with our ignorance about it and the need for its continued and intensive study. He followed with great interest the moving edge of research on that disorder, and perhaps a fitting tribute to him by a biologist would be an examination of recent progress on the biological aspects of schizophrenia.

Chemical hypotheses about insanity were developed by the Hippocratic physicians, but I believe the modern biochemical approach to schizophrenia can be traced to Thudichum. Nearly 100 years ago, this physician and biochemist stated (Thudichum 1884) that many forms of insanity are probably the result of toxic substances fermented within the body, just as the psychoses of alcohol are the result of a toxic substance which is fermented outside. With that theory in mind, he accepted a ten-year research grant by the British Parliament. It is interesting that he did not then turn to the Bedlam Hospital and examine the urine and the blood of patients there; instead he went to the slaughterhouse to obtain cattle brain and spent the ten years under the support of that grant studying the normal composition of that organ. It is very fortunate for us that he did that, because in so doing he laid the foundations of modern neurochemistry, from which some of the answers we seek will eventually come. If he had been less wise, or if the Parliament or the Privy Council had been more insistent that he do

Seymour S. Kety, M.D., Professor of Psychiatry, Harvard Medical School, Director, Psychiatric Research Laboratories, Massachusetts General Hospital, Boston, Mass.

"relevant" research, what kind of research could Thudichum have done on insanity at that time, with the little knowledge that he or anyone else had about biochemistry? What chance would he have had to identify deficits or excesses in substances that were unknown or abnormalities in biochemical processes that were undreamed of? He would have frittered away the public funds and wasted ten years of his life in a premature and futile search.

My first study of schizophrenia took place just twenty-five years ago and consisted in applying to that disorder a newly developed technique for measuring the blood flow and energy metabolism of the brain (Kety & Schmidt 1948). Interestingly enough, the development of that technique was aided by a grant from the Scottish Rite Foundation for Research in Dementia Praecox, which was supporting psychiatric research a decade before the National Institute of Mental Health came into existence. Because that Foundation had supported our research without questioning its relevance, one of the things we wanted to do once we had this method was to apply it to schizophrenia, since it had been suggested that schizophrenia might be the result of an inadequate circulation to the brain or an insufficiency in its utilization of oxygen. So, with Fritz Freyhan at the Delaware State Hospital, and a number of collaborators at the University of Pennsylvania, we undertook a study on a population of thirty schizophrenics displaying various forms of the illness (Kety, Woodford, Harmel, Freyhan, Appel, & Schmidt 1948). We could find no difference from normal in the circulation to the brain or in its oxygen consumption in schizophrenia, and we concluded that it takes just as much oxygen to think an irrational thought as to think a rational one. But, more seriously, we also concluded that if there was a biochemical disturbance in schizophrenia, it probably lay in much more subtle and complex processes than in the overall circulation and energy metabolism of the brain.

Our knowledge of where to look in the biochemical approach to schizophrenia twenty-five years ago was just about as meager as that. We had no promising leads; we had no indications of what these more subtle and complex neurochemical processes might be, and without that, we could not begin to formulate plausible and heuristic hypotheses.

Eight years later, in the Laboratory of Clinical Science at the National Institute of Mental Health, I had the good fortune to be associated with a remarkable group of colleagues—Axelrod, Evarts, Sokoloff, Kopin, Pollin, Cardon, Kies, and LaBrosse, to mention the names of some of the members of the group. There were a few more hypotheses about schizophrenia at that time, representing bold leaps from rather insufficient basic knowledge to the clinical problem. One hypothesis seemed worth examining. It postulated a defect in the metabolism of circulating epinephrine leading to the accumulation of adrenochrome, a supposedly hallucinogenic metabolite (Hoffer 1957). We quickly realized, however, that we didn't know enough about the normal metabolism of epinephrine, let alone its metabolism in schizophrenia. Fortunately, Axelrod was interested in the problem, and, recognizing the importance of the finding of Armstrong, McMillan, and Shaw (1957) that a new metabolite, vanillyl-mandelic acid, was excreted in the urine of patients with pheochromocytoma, he laid out, in a short period of time, the metabolic pathways of epinephrine degradation, demonstrated the importance of a new enzyme, catechol-0-methyl-transferase, and characterized the major metabolites (Axelrod 1959). This made it possible to examine the metabolism of circulating epinephrine in schizophrenia (Labrosse, Mann, & Kety, 1961). We found no abnormalities and no evidence that the schizophrenic patient made adrenochrome from circulating epinephrine. Axelrod's contributions to fundamental knowledge regarding the catecholamine neurotransmitters have already affected our understanding of the drugs useful in treating the major mental illnesses and have stimulated the development of modern neuropharmacology.

There were many other hypotheses at that time centering upon ceruloplasmin or "taraxein" (which was thought to be a modified form of ceruloplasmin), other plasma proteins, and other putative neurotransmitters, but none of these hypotheses had been confirmed. Thus, in 1959, and again some years later, when I had the privilege of reviewing the biochemical aspects of schizophrenia for this association (Kety 1959, 1966), I had to admit that if a young man came to me and asked not for the answer to schizophrenia, not even what the salient biochemical disturbance might be, but simply what aspects of biochemistry he should study if he wanted someday to make a contribution to schizophrenia, I would not have known. I feel differently today. To a considerable extent because of the public support of biomedical, neurobiological, and behavioral research, and the wise philosophy which guided it and its administration, we have learned more about relationships between the brain and behavior in the past twenty-five years than man had learned in all the previous history of civilization. In contrast to the situation twenty-five years ago, the remarkable growth of fundamental knowledge now makes very appropriate and promising the development of heuristic hypotheses regarding the etiology of schizophrenia, which can be rigorously tested with techniques that are now available.

There are two current hypotheses which arose from basic research and which seem capable of tying together more of the observations about schizophrenia and its pharmacology than previous hypotheses. I should like to indicate some of the reasons why these hypotheses have remained viable and why they are quite promising. There is the transmethylation hypothesis which was formulated by Harley-Mason (Osmond & Smythies 1952) and which depended upon Cantoni's demonstration of the important biological process of transmethylation. It was suggested that in schizophrenia there is an accumulation of methylated hallucinogenic substances produced in some way by deviant metabolic pathways or by the inability to degrade normal methylated metabolites that would otherwise be rapidly detoxified. Nicotinamide was proposed as a treatment for schizophrenia on the thesis that it was a methyl acceptor and would drain the methyl groups away from the abnormal pathways. Although this vitamin was reported by the original group and has since been promulgated as an important treatment for schizophrenia (Hoffer & Osmond 1964), that observation has not been confirmed by more than a dozen controlled studies which have been carried out in this country and in Canada (Lipton, Ban, Kane, Levine, Mosher, & Wittenborn 1973). That does not argue against the transmethylation hypothesis, however, since Baldessarini and Kopin (1966) showed some years later that nicotinamide did not reduce S-adenosylmethionine in the brain and could hardly compete with transmethylation there.

On the other hand, methionine administration, which tested the hypothesis at another point, on the assumption that it would increase brain levels of S-adenosylmethionine and favor transmethylation, was employed in our laboratory (Pollin, Cardon, & Kety 1961) and in several subsequent studies (cited by Antun, Burnett, Cooper, Daly, Smythies, & Zealley 1971), and was found to produce an exacerbation of psychosis in a significant fraction of schizophrenics. Although that was compatible with the transmethylation hypothesis, it could not be taken as evidence in its favor because of the several alternative mechanisms by which methionine could produce a psychotic exacerbation. Nor was it ever conclusively shown that the methionine aggravation is, in fact, an aggravation of the schizophrenic process rather than a toxic or reactive psychosis superimposed upon it. On the other hand, methionine produces no suggestion of a psychosis in normal volunteers who have received it, so that the finding appears to have more than a tangential relationship to schizophrenia.

Equally interesting is the enzyme first discovered by Axelrod (1961) in the lung, then in the brain by Mandell and Morgan (1971) and confirmed in the brain by Saavedra and Axelrod (1972), capable of methylating normal indoleamines to hallucinogenic substances. Thus, the enzyme will convert tryptamine to dimethyltryptamine, a powerful hallucinogenic substance, and there have been reports that during administration of a methyl-donor, schizophrenic patients who show exacerbation of psychosis also excrete increased quantities of the hallucinogenic dimethyltryptamine in their urine (Narasimhachari, Heller, Spaide, Haskovec, Fujimori, Tabushi, & Himwich 1971).

There are also the observations of Friedhoff and Van Winkle (1962) of a methylated congener of dopamine, dimethoxyphenylethylamine, closely related to mescaline, in the urine of schizophrenics. Despite considerable controversy, this still remains a provocative finding that is being explored and extended with appropriate analytical techniques. Its specificity for schizophrenia awaits independent confirmation.

I would like to turn now to a second hypothesis which is somehow even more attractive than transmethylation, because it can account for more of the schizophrenic syndrome than the presence of hallucinations. The two hypotheses are not necessarily mutually exclusive, and one can speculate about their relationship with each other. The second hypothesis which is being examined at a number of laboratories postulates that disturbances in central catecholamine synapses may account for the crucial vulnerability of the schizophrenic and for many of his symptoms. This hypothesis is assuming increased importance because of a remarkable series of convergences which strike me as being more than coincidences.

Some twenty years ago, when Paul Hoch gave a talk on schizophrenia at the National Institute of Mental Health, I asked him, attempting to get a clue to the biological mechanisms which might be involved in schizophrenia, what organic psychosis in all of his experience most closely resembled schizophrenia. He answered without hesitation, "the psychosis induced by amphetamine." His experience has been amply confirmed. Much more than LSD, or any of the hallucinogens, amphetamine has the ability to produce in most individuals who receive it in sufficient doses, a toxic psychosis difficult to distinguish from a paranoid schizophrenic reaction. For this we have the most recent testimony of a group (Angrist & Gershon 1971) who have studied this phenomenon extensively. It is interesting that amphetamine has a number of actions, but most of these actions involve the catecholamine system in the brain. Amphetamine releases dopamine and norepinephrine at catecholamine-containing nerve endings and in other ways potentiates the action of these putative transmitters at their synapses in the brain (Glowinski & Axelrod 1965). Here we have the first convergence: amphetamine psychosis (which resembles schizophrenia) may be produced by an activation of catecholamine synapses in the brain. More recently we have had the stimulating neuropharmacological work of Snyder and his colleagues (Taylor & Snyder 1971; Snyder 1972) and the studies of Randrup and Munkvad (1970) to suggest that the psychosis and the stereotyped behavior which is produced by amphetamine in man or animals is likely to be related to its effect at dopamine rather than norepinephrine receptors.

There is also the independent body of knowledge stemming from the discovery of chlorpromazine and the succeeding antipsychotic drugs and dealing with their mechanism of action. There was first the important clinical observation that both the phenothiazines and the butyrophenones, which were discovered by serendipity and quite independently, are both effective against schizophrenia, although they are not related chemically, and both produce symptoms of Parkinson's disease as important side effects. I remember the clinicians who were among the first to use the phenothiazines fifteen to twenty years ago saying that unless one got the beginning of Parkinsonian symptoms, one was probably not giving enough of the drug to produce the greatest therapeutic benefit in schizophrenia.

That important observation became the source of some heuristic speculation about schizophrenia, once our body of fundamental information permitted a better understanding of Parkinson's disease. That, in turn, required the discovery of the fluorescent properties of certain catecholamine-conjugated products, the development of suitable histofluorescent techniques, their application to the brain (Hillarp, Fuxe, & Dahlström 1966), the identification of a dopamine-containing nigrostriatal pathway (Anden, Fuxe, Hamberger, & Hökfelt 1966), and eventually the demonstration (Hornykiewicz 1963) that Parkinson's disease was associated with a deficiency of dopamine in the striatum. That information, coupled with the accumulated knowledge of the biosynthesis of dopamine, led to the effective treatment of Parkinson's disease with L-dopa. It also led to an understanding of the mechanism of action of the antipsychotic drugs and, perhaps, to an elucidation of some of the biological processes at fault in schizophrenia.

In 1963 Carlsson suggested that the phenothiazines produced a blockade of dopamine receptors in the brain. Several laboratories have now (Carlsson & Lindqvist 1963; Anden, Roos, & Verdinius 1964; Nybäck, Borzecki, & Sedvall 1968) adduced evidence that this was so for the phenothiazines and butyrophenones, although the explanation of their therapeutic effects is still incomplete (Matthysse 1973).

Meanwhile, more clinical information was being acquired about the therapeutic effects of these drugs which were at first thought to be merely "chemical straight jackets," suppressing aggressive or disturbed behavior. Davis (1965), examining the results of a large number of double-blind studies on the phenothiazine drugs in schizophrenia, noted that the most prominent effect of these drugs was upon the cardinal features of schizophrenia as they were described by Kraepelin (i.e., on thought disorder, blunted affect, withdrawal, and autistic behavior), which suggest that these drugs are providing more than symptomatic relief, that they may, in fact, be acting more specifically on biological processes in schizophrenia.

Thus, we have seen in the past ten years a series of interesting clinical observations and the accumulating of basic information converging on the activity of dopamine and certain of its synapses in the brain to explain both the psychotic, stereotyped behavior induced by amphetamine and the antipsychotic therapeutic effects of the phenothiazines and butyrophenones. This has suggested to more than one observer that dopamine and its pathways in the brain must be playing a special role in schizophrenia. In fact, the alternative to that inference is that it is a series of coincidences that apomorphine, which produces stereotyped behavior, and the amphetamines, which simulate schizophrenia, also stimulate dopamine receptors in the brain, while the phenothiazines and butyrophenones discovered independently to have important therapeutic effects in schizophrenia and to prevent drug-induced stereotypy, also block dopamine receptors.

Although I feel that these observations are unlikely to be mere coincidences and that dopamine must be playing an important role in schizophrenia, I also feel that dopamine is only part of the story. We must avoid premature closure and keep our thinking open to other neurotransmitters and additional pathways which may be involved. Thus, one may account for the stereotyped behavior of schizophrenia on the basis of increased activity of dopamine receptors, but there are other features of schizophrenia responsive to the antipsychotic drugs which no known action on dopamine receptors can explain. I am thinking of the flatness of affect, the anhedonia, withdrawal and autism of schizophrenia. Some of these manifestations are suggestive of behavioral changes produced in animals by depletion or blockade of norepinephrine.

We are far from understanding the precise behavioral role of that transmitter at its central synapses, but there is considerable evidence to support its involvement in the mediation of exploratory and affective behavior, in motivation and in mood (Segal & Mandell 1970; Stein 1964; Slangen & Miller 1969; Redmond, Maas, Kling, & Dekirmenjean 1971). Monkeys which have received alpha-methyl-paratyrosine, which quite specifically blocks catecholamine synapses, make fairly good animal models of the withdrawal, absence of affect, lack of motivation and even catatonia of the schizophrenic.

It would seem that a defect which would result in an overactivity of dopamine and an underactivity of norepinephrine would account for more of the manifestations of schizophrenia than a change in dopamine activity alone. This focuses our attention immediately on dopamine-beta-hydroxylase (DBH), the enzyme responsible for the conversion of dopamine to norepinephrine. Is it possible that this enzyme operates as a switch in altering the ratio of dopamine to norepinephrine which is produced at certain synapses?

There is a precedent for such an enzyme effect which was discovered by Wurtman and Axelrod (1966) involving an enzyme in the adrenal medulla, phenylethanolamine-N-methyltransferase, which converts norepinephrine to epinephrine. This enzyme appears to play an important role in regulating the ratio of epinephrine to norepinephrine, which is secreted by that gland, and, interestingly enough, the corticosteroids secreted by the adrenal cortex and passing by way of a special portal circulation to the medulla are able to affect the ratio of these two catecholamines by action on the enzyme. Now, if DBH were playing a similar role in the brain, regulating the ratio of norepinephrine to dopamine, a defect in that enzyme or its regulation could produce the alteration in that ratio which is suggested by the manifestations of schizophrenia. Thoenen and associates (Thoenen, Haefeley, Gey, & Hürlimann 1965) have shown that at a peripheral noradrenergic synapse, dopamine is released in a perceptible amount if DBH is inhibited. It is not impossible that noradrenergic endings in the brain are capable of releasing dopamine or norepinephrine, transmitters with different effects, their ratio depending on the DBH activity there and the factors which affect it.

There are some observations on schizophrenia which are compatible with that notion. Antabuse has been reported to produce psychotic reaction in some schizophrenics who receive it (Heath, Nesselhof, Bishop, & Byers 1965). Dopamine-beta-hydroxylase is one of the enzymes that is inhibited by antabuse. Although L-dopa in large doses can produce psychotic side effects, one group has reported that, given in moderate dosage to schizophrenics who are being treated with phenothiazines, it significantly enhanced the therapeutic effect (Inanoga, Inoue, Tachibana, Oshima, & Kotorii 1972). Since the phenothiazine blockade of dopamine receptors is much greater than that on norepinephrine receptors (Neff & Costa 1966), L-dopa administration could result in greater norepinephrine effects if given in conjunction with phenothiazines. The latter drugs, per se, may enhance norepinephrine effects by stimulating catecholamine synthesis via a feedback mechanism activated by a blockade of dopamine receptors, but leaving the norepinephrine receptors relatively free to respond to the increased norepinephrine. Pimozide, a new antipsychotic agent which blocks only dopamine receptors, increases mood and motivation in chronic schizophrenics. This hypothetical mechanism would help to explain such effects.

The most recent and interesting observation pertinent to this hypothesis was made by Wise and Stein (1973) who found a 40 percent reduction in DBH activity in the brains of schizophrenics post mortem as compared to nonschizophrenic controls. A large number of artifacts can occur in such studies and these authors were aware of and took great pains to rule out the effects of age, post-mortem degradation, and chronic phenothiazine administration. Another compelling and well-controlled finding which is free of the difficulties inherent in post-mortem studies is the observation by Wyatt and his collaborators (Wyatt, Murphy, Belmaker, Donnelly, Cohen, & Pollin 1973) at the National Institute of Mental Health, that monoamine oxidase is significantly reduced in the platelets of schizophrenics and equally in the schizophrenic and nonschizophrenic members of monozygotic twin pairs discordant for schizophrenia. The authors suggest that monoamine oxidase may be a genetic marker for schizophrenia. If the reduction in monoamine oxidase also occurs in the brain, it could result in various disturbances in central amine metabolism, including increased dopamine and serotonin activity or the accumulation of hypothetical methylated and hallucinogenic amines.

It is clearly too early to put all of these observations together into a single definitive hypothesis that would "explain" schizophrenia. Nevertheless, I have the feeling, which I have never had before, that we are beginning to see the light at the end of the tunnel. There are powerful new techniques, such as mass fragmentography, more specific pharmacological agents, precursors and enzyme inhibitors. There are new immunochemical techniques for identifying and assaying enzymes and specifying their regional localization (Hartman, Zide, & Udenfriend 1972). There are the techniques of cell culture and the ingenious endocrine window which Sachar has opened for inferring aminergic activity in the human brain from hypothalamically released endocrine responses (Sachar, Finkelstein, & Hellman 1971).

The recent, more conclusive and definitive genetic evidence, such as that reported in the chapters of part III of this volume,* gives ample testimony that schizophrenia is more than a myth and a strong justification for seekng the biological processes required for the expression of the genetic components which are clearly involved in schizophrenic illness. Moreover, in contrast to the situation one or two decades ago, we now have some fair ideas of where to look for the biological substrates on which life experience builds. The foundations which the National Institute of Mental Health helped to build up during the past twenty-five years now appear to be capable of bearing weight. The store of fundamental information acquired during that time provides the basis for a number of heuristic and plausible hypotheses which do not attempt to solve the problem all at once, but indicate important stepping stones to that goal (Kety & Matthysse 1972). We can also see large areas of behavior and biology which remain to be further elucidated and joined. During that time, research has become a powerful ally of psychiatry (Kety 1961), substituting inquiry for dogma, scientific examination for impressionistic thinking. A cohort of psychiatrists has been trained, skilled in clinical investigation and fundamental research. Never has the time been more propitious or progress more promising. We can assure this progress by a prudent distribution of our resources, by investing more than a fraction of one percent of the cost of mental illness in clinical and basic research, and in continued research training. This we will do if we are not hampered by a penny-wise, pound-foolish fiscal policy, or misled by those who would solve the problem of mental illness by denying its existence.

References

Anden, N.-E., Roos, B.-E., & Verdinius, B. Effects of chlorpromazine, haloperidol and reserpine on the level of phenolic acids in the rabbit corpus striatum. *Life Sciences*, 1964, 3, 149–158.

*Kidd (1975); Kety, Rosenthal, Wender, Schulsinger, and Jacobsen (1975); Shields, Heston, and Gottesman (1975); and Rosenthal (1975).

Anden, N.-E., Fuxe, K., Hamberger, B., & Hökfelt, T. A quantitative study on the nigro-neostriatal dopamine neuron system in rat. *Acta Physiologica Scandinavica*, 1966, **67**, 306–312.

Angrist, B. M., & Gershon, S. A pilot study of pathogenic mechanisms in amphetamine psychosis utilizing differential effects of D- and L-amphetamine. *Pharmakopsychiatrie*, 1971, **4**, 64–75.

Antun, F. T., Burnett, G. B., Cooper, A. J., Daly, R. J., Smythies, J. R., & Zealley, A. K. The effects of l-methionine (without MAOI) in schizophrenia. *Journal of Psychiatric Research*, 1971, **8**, 63–71.

Armstrong, M. D., McMillan, A., & Shaw, K. N. F. 3-methoxy-4-hydroxy-D-mandelic acid, a urinary metabolite of norepinephrine. *Biochimica et Biophysica Acta*, 1957, **25**, 422–423.

Axelrod, J. The metabolism of catecholamines *in vivo* and *in vitro*. *Pharmacological Reviews*, 1959, **11**, 402–408.

Axelrod, J. Enzymatic formation of psychotomimetic metabolites from normally occurring compounds. *Science*, 1961, **134**, 343–344.

Baldessarini, R. J., & Kopin, I. S-adenosylmethionine in brain and other tissues. *Journal of Neurochemistry*, 1966, **13**, 764–777.

Carlsson, A., & Lindqvist, M. Effect of chlorpromazine or haloperidol on formation of 3-methoxytyramine and normetanephrine in mouse brain. *Acta Pharmacologica et Toxicologica*, 1963, **20**, 140–144.

Davis, J. M. Efficacy of the tranquilizing and antidepressant drugs. *Archives of General Psychiatry*, 1965, **13**, 552–572.

Friedhoff, A. J., & Van Winkle, E. Characteristics of an amine found in urine of schizophrenic patients. *Journal of Nervous and Mental Disease*, 1962, **135**, 550–555.

Glowinski, J., & Axelrod, J. Effects of drugs on the uptake, release and metabolism of H³-norepinephrine in the rat brain. *Journal of Pharmacology*, 1965, **149**, 43–49.

Hartman, B. K., Zide, D., & Udenfriend, S. The use of dopamine-beta-hydroxylase as a marker for the central noradrenergic system in rat brain. *Proceedings of the National Academy of Sciences* (Washington, D.C.), 1972, **69**, 2722–2726.

Heath, R. G., Nesselhof, W., Bishop, M. P., & Byers, L. N. Behavioral and metabolic changes associated with administration of tetra-ethylthiuram desulfide. *Diseases of the Nervous System*, 1965, **26**, 99.

Hillarp, N.-A., Fuxe, K., & Dahlström, A. Demonstration and mapping of central neurons containing dopamine, noradrenaline, and 5-hydroxytryptamine and their reactions to psychopharmaca. *Pharmacological Reviews*, 1966, **18**, 727–741.

Hoffer, A. Epinephrine derivatives as potential schizophrenic factors. *Journal of Clinical and Experimental Psychopathology*, 1957, **18**, 27.

Hoffer, A., & Osmond, H. Treatment of schizophrenia with nicotinic acid. *Acta Psychiatrica Scandinavica*, 1964, **40**, 171–189.

Hornykiewicz, O. Die topische Lokalisation und das Verhalten von Noradrenalin und Dopamin in der substantia negra des normalen und parkinsonkranken Menschen. *Wiener klinische Wochenschrift*, 1963, **75**, 309–312.

Inanoga, K., Inoue, K., Tachibana, H., Oshima, M., & Kotorii, T. Effect of L-dopa in schizophrenia. *Folia Psychiatrica Neurologica Japonica*, 1972, **26**, 145–157.

Kety, S. S. Biochemical theories of schizophrenia. A two-part critical review of current theories and of the evidence used to support them. *Science*, 1959, **129**, 1528-1532; 1590-1596.

Kety, S. S. The academic lecture: the heuristic aspect of psychiatry. *American Journal of Psychiatry*, 1961, **118**, 385-397.

Kety, S. S. Current biochemical research in schizophrenia. In. P. H. Hoch & J. Zubin (Eds.), *Psychopathology of schizophrenia*. New York: Grune & Stratton, 1966, pp. 225-232.

Kety, S. S., & Matthysse, S. (Eds.) Prospects for Research on Schizophrenia. *Neurosciences Research Program Bulletin*, 1972, **10**, 449-495.

Kety, S. S., Rosenthal, D., Wender, P. H., Schulsinger, F., and Jacobsen, P. Mental illness in the biological and adoptive families of adopted individuals who have become schizophrenic: a preliminary report based on psychiatric interviews. In R. R. Fieve, D. Rosenthal, & H. Brill (Eds.), *Genetic research in psychiatry*. Baltimore: The Johns Hopkins University Press, 1975, pp. 147-66.

Kety, S. S., & Schmidt, C. F. Nitrous oxide method for the quantitative determination of cerebral blood flow in man: theory, procedure and normal values. *Journal of Clinical Investigation*, 1948, **27**, 476-483.

Kety, S. S., Woodford, R. B., Harmel, M. H., Freyhan, F. A., Appel, K. E., & Schmidt, C. F. Cerebral blood flow and metabolism in schizophrenia. The effects of barbituate semi-narcosis, insulin coma and electroshock. *American Journal of Psychiatry*, 1948, **104**, 765-770.

Kidd, K. K. On the possible magnitudes of selective forces maintaining schizophrenia in the population. In R. R. Fieve, D. Rosenthal, & H. Brill (Eds.), *Genetic research in psychiatry*. Baltimore: The Johns Hopkins University Press, 1975, pp. 135-46.

LaBrosse, E. H., Mann, J. D., & Kety, S. S. The physiological and psychological effects of intravenously administered epinephrine and its metabolism in normal and schizophrenic men. III. Metabolism of 7-H^3-epinephrine as determined in studies on blood and urine. *Journal of Psychiatric Research*, 1961, **1**, 68-75.

Lipton, M. E., Ban, T. A., Kane, F. J., Levine, J., Mosher, L. R., & Wittenborn, R. Report of the American Psychiatric Association Task Force on Vitamin Therapy and Psychiatry. Washington: American Psychiatric Association, 1973.

Mandell, A. J., & Morgan, M. Indole(ethyl)amine N-methyl-transferase in human brain. *Nature. New Biology*, 1971, **230**, 85-87.

Matthysse, S. The state of the evidence. In S. S. Kety & S. Matthysse (Eds.), *Symposium on catecholamines and their enzymes in the neuropathology of schizophrenia*. London: Pergamon, in press.

Narasimhachari, N., Heller, B., Spaide, J., Haskovec, L., Fujimori, M., Tabushi, K., & Himwich, H. E. Urinary studies of schizophrenics and controls. *Biological Psychiatry*, 1971, **3**, 9-20.

Neff, N. H., & Costa, E. Effect of tricyclic antidepressants and chlorpromazine on brain catecholamine synthesis. Proceedings of the First International Symposium on Antidepressant Drugs. *Excerpta Medica. International Congress Series*, 1966, No. 122, Milan, pp. 28-34.

Nybäck, H., Borzecki, Z., & Sedvall, G. Accumulation and disappearance of catecholamines formed from tyrosine-^{14}C in mouse brain. *European Journal of Pharmacology*, 1968, **4**, 395-403.

Osmond, H., & Smythies, J. Schizophrenia: a new approach. *Journal of Mental Science*, 1952, **98**, 309–315.

Pollin, W., Cardon, P. V., & Kety, S. S. Effects of amino acid feedings in schizophrenic patients treated with iproniazid. *Science*, 1961, **133**, 104–105.

Randrup, A., & Munkvad, I. Biochemical, anatomical and psychological investigations of stereotyped behavior induced by amphetamines. In E. Costa & S. Garattini (Eds.), *Amphetamines and related compounds*. New York: Raven, 1970, pp. 695–713.

Redmond, D. E., Jr., Maas, J. W., Kling, A., & Dekirmenjean, H. Changes in primate social behavior after treatment with alpha-methyl-paratyrosine. *Psychosomatic Medicine*, 1971, **33**, 97–113.

Rosenthal, D. Discussion: the concept of subschizophrenic disorders. In R. R. Fieve, D. Rosenthal, & H. Brill (Eds.), *Genetic studies in psychiatry*. Baltimore: The Johns Hopkins University Press, 1975, pp. 199–208.

Saavedra, J. M., & Axelrod, J. Psychotomimetic N-methylated tryptamines: formation in brain *in vivo* and *in vitro*. *Science*, 1972, **175**, 1365.

Sachar, E. J., Finkelstein, J., & Hellman, L. Growth hormone and responses in depressive illness: Response to insulin tolerance test. *Archives of General Psychiatry*, 1971, **24**, 263–264.

Segall, D. S., & Mandell, A. J. Behavioral activation of rats during intraventricular infusion of norepinephrine. *Proceedings of the National Academy of Sciences* (Washington, D.C.), 1970, **66**, 289–293.

Shields, J., Heston, L. L., & Gottesman, I. I. Schizophrenia and the schizoid: the problem for genetic analysis. In R. R. Fieve, D. Rosenthal, & H. Brill (Eds.), *Genetic research in psychiatry*. Baltimore: The Johns Hopkins University Press, 1975, pp. 167–96.

Slangen, J. L., & Miller, N. E. Pharmacological tests for the function of hypothalamic norepinephrine in eating behavior. *Physiology and Behavior*, 1969, **4**, 543–552.

Snyder, S. H. Catecholamines in the brain as mediators of amphetamine psychosis. *Archives of General Psychiatry*, 1972, **27**, 169–179.

Stein, L. Self-stimulation of the brain and the central stimulant action of amphetamine. *Federation Proceedings*, 1964, **23**, 836–850.

Taylor, K. M., & Snyder, S. H. Differential effects of d- and l-amphetamine on behavior and on catecholamine in dopamine and norepinephrine containing neurons of rat brain. *Brain Research*, 1971, **28**, 295–309.

Thoenen, H., Haefely, W., Gey, K. F., & Hürlimann, A. Diminished effects of sympathetic nerve stimulation in cats pretreated with disulfiram; liberation of dopamine as sympathetic transmitter. *Life Sciences*, 1965, **4**, 2033–2038.

Thudichum, J. L. W. *A treatise on the chemical constitution of the brain based throughout upon original researches*. London: Bailliere, Tindall and Cox, 1884.

Wise, C. D., & Stein, L. Dopamine-beta-hydroxylase deficits in the brains of schizophrenic patients. *Science*, 1973, **181**, 344–347.

Wurtman, R. J., & Axelrod, J. Control of enzymatic synthesis of adrenaline in the adrenal medulla by adrenal cortical steroids. *Journal of Biological Chemistry*, 1966, **241**, 2301–2305.

Wyatt, R. J., Murphy, D. L., Belmaker, R., Donnelly, C., Cohen, S., & Pollin, W. Reduced monoamine oxidase activity in platelets: a possible genetic marker for vulnerability to schizophrenia. *Science*, 1973, **179**, 916–918.

3 GENETIC ASPECTS OF CATECHOLAMINE SYNTHESIS

Jack D. Barchas, Roland D. Ciaranello,
Seymour Kessler, and David A. Hamburg

A wide variety of studies have demonstrated important relations between catecholamines and behavior leading to the recognition that catecholamines may be of vital importance in normal behaviors as well as in severe psychiatric disorders (reviewed by Barchas, Stolk, Ciaranello, & Hamburg 1971; Barchas, Ciaranello, Stolk, Brodie, & Hamburg 1972a). In light of the potential importance of the catecholamine system, the possibility that there may be genetic controls over the regulation of catecholamine synthesis becomes germane to behavioral considerations. In this paper, we would like to bring together our data relevant to genetic variation in the enzymes involved in the synthesis of the catecholamines using animal models. We will demonstrate that (1) strain differences exist in the activity of the enzymes involved in the synthesis of the catecholamines in mice, (2) that this variation has a genetic basis, and (3) that differential regulatory mechanisms of the enzymes involved in catecholamine formation exist in different strains, and, presumably, that this variation also has a genetic basis. We will present these findings and then discuss the potential implications that genetic regulation of this important neurochemical system holds for severe disorders, with particular attention to depression, an illness to which catecholamines have been particularly linked.

The possibility of genetic variation in catecholamine synthesizing activity is suggested by several lines of evidence. In animals, strain and subline differences have been reported in the amounts of biogenic amines

Jack D. Barchas, M.D., Roland D. Ciaranello, M.D., Seymour Kessler, Ph.D., and David A. Hamburg, M.D., Department of Psychiatry, Stanford University School of Medicine, Stanford, California 94305.

Our work has been made possible by generous grants from the National Institute of Mental Health (PHS Grant MH 23861), the Office of Naval Research, The Grant Foundation, and the Commonwealth Fund.

We would like to acknowledge the important involvement of John Shire, Jeffrey Dornbusch, and Rebecca Barchas in our studies and to thank Elizabeth Erdelyi, Pamela Angwin, and Humberto Garcia for their excellent technical assistance. We appreciate the secretarial assistance of Lynn Hassler, Florence Parma, and Rosemary Schmele in the preparation of this manuscript. We are particularly thankful to Dr. Ronald Fieve for his encouragement, and to David Zubin for his editorial assistance.

in brain regions of mice (Maas 1962, 1963; Sudak & Maas 1964*a*; Schlesinger, Boggan, & Freedman 1965; Karczmar & Scudder 1967) and rats (Sudak & Maas 1964*b*; Miller, Cox, & Maickel 1968) and in the utilization and uptake of cardiac norepinephrine in mice (Page, Kessler, & Vesell 1970). In humans, a variety of forms of pheochromocytoma have been shown to be associated with familial factors (Rimoin & Schimke 1971, pp. 251–57). The known genetic variation in adrenocortical function (Badr & Spickett 1965; Stempfel & Tomkins 1966; Hamburg & Kessler 1967) and in adrenocortical and adrenomedullary structure (Chai & Dickie 1966; Shire 1970) further suggested to us that the search for genetic variation in the enzymes involved in catecholamine biosynthesis would be fruitful.

The use of inbred mouse strains is particularly advantageous in studies of genetic variation and provides a powerful tool for subsequent behavioral and genetic analysis. Each inbred strain represents a distinct genotypic constellation; differences found between strains, when environmental conditions are held constant, suggest that genetic variation may be present. By appropriate matings, the nature of this variation can be elucidated.

We have investigated catecholamine synthesis in both the brain and the adrenal gland. Studies utilizing the adrenal have been particularly helpful because the mechanisms involved in regulation of enzymatic activity in the adrenal have been determined to a greater degree than in other areas, thereby permitting more detailed evaluation of strain and genetic differences. Thus, the use of inbred strains of mice for investigation of brain and adrenal processes has permitted studies of the regulation of catecholamine synthesis which would have been extremely difficult or impossible in other species, and provides an ideal model system for investigation.

Aspects of the Synthesis of Catecholamines

Catecholamines are a class of compounds which include dopamine, norepinephrine, and epinephrine. The catecholamines are involved in two completely different functional processes which are relevant to our concerns. The first process involves catecholamines formed and released by the adrenal gland. Epinephrine (adrenaline) and norepinephrine (noradrenaline) are the two principal members of this class. Compounds released from the adrenal affect other areas of the body, including the brain, although only small amounts can enter the brain due to a barrier (blood-brain barrier) which restricts their entry. In the second role, catecholamines are viewed as transmitter agents between adjacent nerve cells in the brain. Thus, catecholamines formed by the adrenal gland act as circulating hormones, while those formed in the brain act as transmitter agents. Both adrenal and brain must be viewed as separate, extremely active catecholamine-synthesizing compartments with little or no cross-

supply between the two. The catecholamines and the route of their formation are shown in Figure 1; the biochemical pathways involved have been
reviewed by Molinoff and Axelrod (1971).

The first step in the formation of the catecholamines involves the
enzyme tyrosine hydroxylase (TH), which converts the amino acid tyrosine
to DOPA (dihydroxyphenylalanine). From the work of the Udenfriend
group and of others, tyrosine hydroxylase is generally considered to be the
rate-limiting step in catecholamine formation and thus assumes an important role in regulating the levels and activity of catecholamines in
various tissues (Spector, Gordon, Sjoerdsma, & Udenfriend 1967; Nagatsu,
Levitt, & Udenfriend 1964; Sedvall, Weise, & Kopin 1968).

It has become clear from the work of several investigators (Euler 1962;
Gordon, Spector, Sjoerdsma, & Udenfriend 1966; Alousi & Weiner 1966;
Roth, Stjärne, & von Euler 1966; Sedvall & Kopin 1967) that when the nerves
to adrenergically innervated tissues are stimulated the activity of tyrosine
hydroxylase catecholamines is increased. The activity of the enzyme can
increase acutely two- to threefold, thereby allowing for maintenance of
catecholamine levels in situations of heavy use. Only under extreme stress
are catecholamine levels decreased; under such conditions, even a brisk
increase in synthesis cannot keep pace with the vigorous utilization. Some
evidence suggests that normally the levels of catecholamine in the tissue
may regulate the activity of tyrosine hydroxylase; when the level of catecholamine begins to decrease, the enzyme activity is increased. This
process constitutes feedback inhibition (Spector et al. 1967; Patrick &
Barchas, 1974). Another mechanism may come into play with long-
term stimulation in which the actual number of tyrosine hydroxylase
molecules increases (Weiner & Rabadjija 1968). Other important aspects
of mechanisms for regulation of tyrosine hydroxylase in the adrenal gland
include neuronally mediated induction of the enzyme (Mueller, Thoenen,

Fig. 1. Metabolic Pathway of Catecholamine Synthesis.

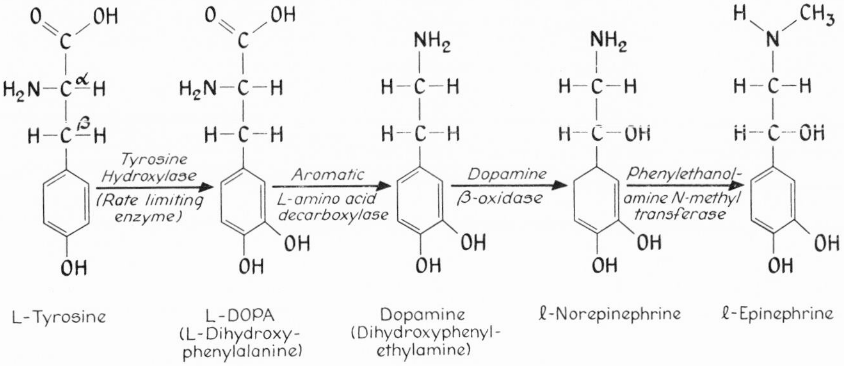

& Axelrod 1969; Thoenen, Mueller, & Axelrod 1969). Activity may also be influenced directly by ACTH (Mueller et al. 1969).

The product of tyrosine hydroxylation, dihydroxyphenylalanine (DOPA), is rapidly decarboxylated to dopamine by the enzyme aromatic L-amino acid decarboxylase. Normally, dopamine is found in high concentrations primarily in the basal ganglia of the brain. Dopamine has important roles in motor function and may also have a role in emotional behaviors and endocrine function, as well as function in the periphery in relation to the sympathetic innervation of the blood vessels. Although it is the second enzyme involved in the synthesis of catecholamines, the decarboxylase has thus far not shown substantial variation with physiological or psychological states.

The third enzymic step in the production of the catecholamines involves the addition of an hydroxyl (OH) group to the β-carbon of dopamine, thereby forming norepinephrine. This conversion of dopamine to norepinephrine is accomplished by the enzyme dopamine-β-oxidase (DβO) (Friedman & Kaufman 1965). Norepinephrine and dopamine are generally thought to be the predominant transmitters in the peripheral adrenergic nervous system. This is the system that contributes the bulk of the catecholamines that appear in the urine. Changes in urinary catecholamine levels or catecholamine metabolite content as a response to stress of other forms of behavior reflect alterations in peripheral rather than central nervous system catecholamine concentration. These alterations are brought about as a consequence of changes in impulse flow to adrenergically innervated organs, and to secretion into the blood of the catecholamines stored in the adrenal medulla. The brain contains enzymatic machinery for forming norepinephrine where it may serve as a transmitter in certain areas of the brain (Vogt 1954) and indeed the brain may form more catecholamines per unit time than the adrenal medulla, a tissue specialized in the formation of catecholamines.

The regulation of DβO activity, a key step in norepinephrine formation in adrenergic nervous tissue, is being investigated at present in many laboratories. There are inhibitors of the enzyme present in tissues where the enzyme is found. One could postulate that DβO, in certain circumstances, acts as the rate-limiting step in norepinephrine biosynthesis, rather than TH. Among studies of the regulation of this enzyme are those from the laboratory of Axelrod and his co-workers. Molinoff, Brimijoin, Weinshilboum, and Axelrod (1970) demonstrated that neural activity plays a role in the maintenance of enzyme activity in sympathetic ganglia. Weinshilboum and Axelrod (1970) have demonstrated that the adrenal enzyme is reduced markedly in the hypophysectomized rat; activity can be restored by ACTH but not by glucocorticoids. These investigators have interpreted their results to mean that the enzyme is under complex endocrine control with ACTH possibly being the primary regulator of enzyme activity. Although little is known about regulation of this enzymatic step, it could prove of great interest in terms of behavior (Weinshilboum, Raymond, Elueback, & Weidman, 1973).

The final synthetic enzyme, phenylethanolamine N-methyl transferase (PNMT), converts norepinephrine to epinephrine. Enzyme activity is high only in the adrenal medulla (Axelrod 1962), although low levels of epinephrine formation have been found in mammalian brain (Barchas, Ciaranello, & Steinman 1969). The conversion of norepinephrine to epinephrine by PNMT has been shown to be altered by a number of factors, including adrenal steroids (Molinoff & Axelrod 1971; Wurtman & Axelrod 1966). The regulation of the formation of epinephrine from norepinephrine by the enzyme PNMT in the adrenal gland appears to involve factors including both the adrenal cortex and the adrenal medulla. The adrenal gland with its distinct outer cortex and inner medulla is actually two glands in one. The cortex secretes steroid hormones and plays an essential role in the stress response. Adrenal steroid output is mediated through the pituitary gland and is stimulated by a pituitary hormone, adrenocorticotropic hormone (ACTH). On the other hand, the adrenal medulla secretes hormones during stress—epinephrine and norepinephrine. Secretion from the adrenal is controlled via peripheral and central nervous system activity.

Several avenues of work (Coupland 1953; Wurtman & Axelrod 1966) have demonstrated that adrenal steroids are essential in maintaining the levels of PNMT, and that in the absence of adrenal steroids, the level of PNMT is markedly reduced. The work of Wurtman and Axelrod (1966) demonstrated that hypophysectomy causes a decrease of PNMT activity in adrenal tissue to about 10 percent of normal. Activity could be restored by administering cortisol, ACTH, or dexamethasone. When a variety of other pituitary hormones were tested, only ACTH caused this effect on PNMT activity. The stimulation of PNMT in the hypophysectomized animal was shown to proceed by an increase in enzyme synthesis. When protein synthesis inhibitors were administered with the glucocorticoid, there was no detectable rise in enzyme activity. Neither did the addition of glucocorticoids or ACTH to the enzyme *in vitro* increase activity. *In vivo*, however, it was found initially that the enzyme could not be stimulated in the intact animal using ACTH, dexamethasone, or a variety of short-term stresses (Fuller & Hunt 1967).

It now appears that enzyme activity can be elevated in the intact animal, but that its rate of rise proceeds much more slowly than in the hypophysectomized animal. Thus, in the hypophysectomized rat, PNMT activity doubles after 3 to 4 days of dexamethasone or ACTH treatment. Using prolonged physiological stresses, Vernikos-Danellis, Ciaranello, and Barchas (1968) and Ciaranello, Barchas, and Vernikos-Danellis (1969b) demonstrated that PNMT activity doubled in the intact animal over a 20 to 50 day period.

These results suggest that control mechanisms operating in the intact rat subjected to physiological stress are different from the mechanisms which regulate PNMT activity in the hypophysectomized animal treated with ACTH or glucocorticoids. The effects of the relatively slow turnover of PNMT and epinephrine in the intact animal might be to ensure adequate supplies of epinephrine in times of stress and to limit the amount of epinephrine released at any given time, which could have detrimental effects on the organism. The increase in PNMT activity with long-term stress could have special significance in terms of physiological, biochemical, and behavioral aspects of chronic stress.

Studies have demonstrated that PNMT also is under partial neuronal control. When the splanchnic nerves to the adrenal are stimulated by administration of the drug 6-hydroxydopamine, PNMT activity rises slowly (Mueller et al. 1969). This rise can be blocked by denervation, suggesting that the response is neuronally triggered. Hypophysectomy has no effect on the enzyme response to 6-hydroxydopamine, indicating that the pituitary-adrenocortical axis is not involved in this aspect of PNMT regulation (Thoenen, Mueller, & Axelrod 1970). Similarly, these same investigators demonstrated that denervation has no effect on the response of the enzyme to dexamethasone in hypophysectomized animals. These results are interpreted to mean that the neuronal and hormonal controls on the enzyme are mediated via different routes.

Through the various studies which have been performed, it is apparent that the pathway involving catecholamine synthesis involves a number of steps which can be important in the regulation of the final catecholamine products. The regulation of the steps involves a number of variables which further suggest the potential importance of genetic investigations.

Genetic Differences in Levels of Enzymes Synthesizing Catecholamines

Because of the importance of TH and PNMT in the synthesis of the catecholamines, we decided to investigate the activity of those enzymes in different inbred mouse strains. For each of these enzymes there is clear indication of their regulatory importance in determining the levels of the formed catecholamines, and many of the mechanisms involving the enzymes have been determined.

For these studies, mice of various strains were obtained from Jackson Laboratory, Bar Harbor, Maine, except for the C57Bl / Ka which were obtained from the Department of Radiology at Stanford. The animals were in the 20 to 25 gram range and were about 6 to 8 weeks old at the time of the various studies. The animals were usually housed in groups of 6 in a room with a cycle of 13 hours of light and 11 hours of dark for 4 to 5 days prior to the studies. When the comparison of several strains of mice was involved, the pertinent assays were performed on the brains or adrenals on the same day. The assay for PNMT was performed by the method of Axelrod (1962), and the small amount of radioactive methanol formed enzymatically as an interfering product (Ciaranello, Dankers, & Barchas 1972*b*) was removed by the procedure of Deguchi and Barchas (1971). The assay for TH was performed by the method of Levitt (personal communication), and catecholamine levels were determined using the procedure of Barchas, Erdelyi, & Angwin (1972*b*).

The first set of studies involved the comparison of the levels of TH, PNMT, and adrenal catecholamines in five strains of mice (Kessler, Ciaranello, Shire, & Barchas 1971; Ciaranello, Barchas, Kessler, & Barchas 1972*a*). As shown in Figure 2, there are marked differences in the levels of the enzymes in the adrenal in the different strains. The variation in the level of the enzymes between different inbred strains was further evaluated in terms of the levels of the catecholamines in the adrenal. Although significant differences were noted between the strains in the levels of

Fig. 2. PNMT and TH Activity in Inbred Mouse Strains. PNMT activity is expressed as the number of millimicromoles N-[methyl-^{14}C]-phenylethanolamine formed per hour per adrenal. TH activity is expressed as the amount of DOPA-^{14}C formed per hour per adrenal. Mean ($\pm$ S.E.) adrenal PNMT and TH activities in inbred mouse strains (6 animals per strain).

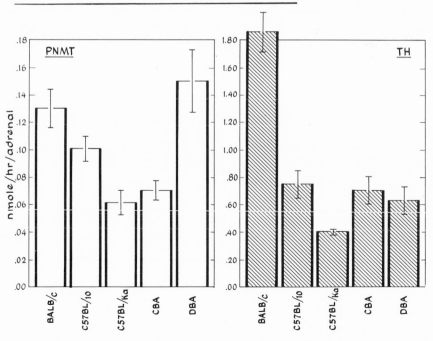

Table 1. Brain Tyrosine Hydroxylase Activities in Inbred Mouse Strains

Strain[b]	Brain TH (nmoles/hr/g)[a]	Brain weight (g)
BALB/cJ	20.09 ± 2.79	0.4456 ± 0.0128
C57B1/Ka	15.66 ± 2.00	0.3439 ± 0.0116
C57B1/10	10.76 ± 0.54	0.4152 ± 0.0090
CBA	10.03 ± 0.53	0.4166 ± 0.0251
DBA/2J	10.98 ± 0.67	0.3532 ± 0.0092

[a] Means ± S.E.
[b] Six animals per group.

adrenal norepinephrine and epinephrine, no obvious relationship between the level of the enzymes and the level of the adrenal catecholamines was found. This is not surprising, since the measured level of adrenal catecholamines results from the operation of many factors, including the rate of synthesis, the rate of release from the adrenal, turnover, storage, etc. Furthermore, no relationship between adrenal weight and relative enzyme activity was found; the results suggest that among the interstrain enzyme differences noted, most or all exist independent of any morphological differences between the adrenal glands of the various strains.

The levels of tyrosine hydroxylase in the brain were determined in the five strains of mice as shown in Table 1. There were marked differences between strains. Analysis of variance showed the presence of significant ($p < .01$) strain differences. The relative rank order of strains was identical for both adrenal and brain TH, with the notable exception of C57B1/Ka, which showed the lowest adrenal TH in combination with a relatively high activity of brain TH. This finding raises the possibility that structurally different forms of this enzyme and/or differences in the rates of turnover exist in the two tissues.

The results presented demonstrate strain differences in the levels of the enzymes involved in catecholamine synthesis. It is important, however, to show that these differences have a genetic basis. Demonstration of strain differences may not be sufficient proof of the presence of genetic variation as differential treatment of pups by their mothers may produce environmentally induced differences. In the few studies concerned with strain differences in levels of catecholamines (e.g., strain differences in levels of brain catecholamines), the demonstration of a genetic basis of the differences was not attempted. Therefore, in a separate series of studies, we crossed several of the strains (Kessler, Ciaranello, Shire, & Barchas 1972) and measured enzyme activity in the resulting progenies. Some of the results are shown in Table 2. From this analysis, a number of interesting points are apparent. Genetic factors play an important role in the determination of the differences between the strains. All the recipro-

Table 2. Activities[a] of Enzymes Involved in Synthesis of Catecholamines

		Brain	Adrenal	
	n	Tyrosine hydroxylase[b]	Tyrosine hydroxylase[c]	Phenylethanolamine N-methyltransferase[d]
Strains:				
BALB/cJ	8	14.6 ± 1.14	5.28 ± 0.81	0.198 ± 0.015
CBA/J	8	9.4 ± 0.66	1.99 ± 0.14	0.124 ± 0.007
C57B1/Ka	8	11.6 ± 0.99	1.40 ± 0.09	0.085 ± 0.008
F_1 mice:				
BALB/cJ × CBA/J	16	12.5 ± 0.66	2.24 ± 0.10	0.119 ± 0.008
CBA/J × C57B1/Ka	16	11.7 ± 0.51	1.82 ± 0.16	0.143 ± 0.010
C57B1/Ka × BALB/cJ	16	10.4 ± 0.70	2.17 ± 0.10	0.118 ± 0.006

[a] Activities expressed as the mean ± S.E.
[b] nmol of [^{14}C] DOPA formed per hr per g of brain tissue.
[c] nmol of [^{14}C] DOPA formed per hr per pair of adrenals.
[d] nmol of N-[^{14}C] methylphenylethanolamine formed per hr per pair of adrenals.

cal crosses involving a given pair of strains yielded similar values; therefore pre- and postnatal biological and cultural maternal effects and also X-linkage could be eliminated as important determinants of the observed differences. In Table 2, the reciprocal hybrid combinations were pooled because they were not significantly different. In the adrenal gland, PNMT and TH activities were correlated over all the genotypic groups studied, suggesting that in these strains the two enzymes may be controlled by the same genetic factor(s), possibly due to close linkage on the same chromosome or to coordinate gene regulation. These possibilities might be further tested by studying whether the correlation between the enzyme activities is maintained or breaks down in segregating generations (F_2 and back-crosses) (Shire 1969).

The pattern of inheritance of brain tyrosine hydroxylase differs from that found for adrenal tyrosine hydroxylase; the gene(s) of C57Bl / Ka are dominant to those of the other two strains. The cross between BALB/cJ and CBA/J produced hybrids with brain tyrosine hydroxylase activity intermediate between the parental strains.

Our findings show that in house mice (*Mus musculus*) genetic variation which affects the enzymes involved in the synthesis of catecholamines exists within the normal range. This variation could result either from differences in the number of enzyme molecules, or from differences in their structure, or both. Axelrod and Vesell (1970) found that adrenal phenyl-ethanolamine N-methyltransferase showed no intraspecific heterogeneity in several species when subjected to starch-block electrophoresis or when its heat stability was examined. However, a broader sampling, beginning with the strains studied here, might reveal either or both enzymes to be structurally heterogeneous within a single species.

The patterns of inheritance of tyrosine hydroxylase activity are different in the adrenal gland and in the brain. Therefore, there must also be differences at the biochemical level in these tissues. It is possible that different enzymes are present in the two tissues. Alternatively, a single kind of protein molecule with tyrosine hydroxylase activity may be present in both adrenal gland and brain, but genetically controlled differences in the mechanisms regulating the synthesis, activation, and degradation of the enzyme may be operative in the two tissues.

CBA/J and C57Bl / Ka appear to carry dominant genes for the adrenal enzymes and for brain tyrosine hydroxylase, respectively. The activities of both adrenal enzymes in the F_1 hybrids resembled the activities in the inbred CBA/J strain, which, relative to the other inbred strains studied here, had intermediate enzyme activities. Similarly, the brain tyrosine hydroxylase activity in all the F_1 hybrids resembled that of the intermediate inbred strain, in this case C57Bl / Ka. Thus, for all three enzyme characters, dominant genes appear to determine intermediate rather than extreme activity—a finding which is of considerable theoretical interest to geneticists. Clearly, variation in the ability to synthesize catecholamines may be an important factor in accounting for individual differences in responses to stress.

Recent studies have contributed further to our understanding of the mechanism of gene control of steady-state levels (Ciaranello & Axelrod 1973; Ciaranello 1973). Two sublines of the BALB/c strain were used in these studies, the BALB/cJ bred at Jackson Laboratories and the BALB/cN, bred at the NIH. These sublines are of particular interest since they were both derived from a common ancestral colony and have been bred independently for over twenty-five years.

The BALB/cJ line has approximately twice the adrenal gland activity of tyrosine hydroxylase, dopamine-β-hydroxylase, and phenylethanolamine N-methyltransferase as the BALB/cN line (Ciaranello, Hoffman, Shire, & Axelrod 1974). Using antisera specific to each enzyme, this difference in activity was shown to be due to differences in the number of molecules of each enzyme. Detailed screening studies failed to disclose any other biochemical difference between the mouse sublines, suggesting that a mutation or mutations had occurred which specifically affected the levels of adrenal TH, DBH, and PNMT.

Analysis of hybrid (F_1) and segregating (F_2 and backcross) progeny showed that the steady-state level of each enzyme was under the control of a single genetic locus. Correlation and phenotype distribution analyses suggested that the phenotypic traits were inherited in association rather than as separate traits, i.e., a given mouse tended to be high, intermediate, or low in all three enzymes. These findings suggested that inheritance of enzyme phenotypes followed the pattern of single-factor inheritance rather than that of three separate genes. Mechanistically, this could occur as a result of linkage of three genes on a single chromosome. Alternatively, a single regulatory gene could be controlling the phenotypic expression of all three enzymes (Ciaranello et al. 1974).

Studies were done to distinguish between these alternatives. The enzymes could have been twice as high in the BALB/cJ line because it was making the enzymes twice as rapidly, or because it was degrading them at half the rate, or some combination of an altered rate of synthesis and degradation. Since precedents exist for each of these alternatives, there is no *a priori* reason for favoring any of them. PNMT was the first enzyme to be examined (Ciaranello and Axelrod 1973; Ciaranello 1973). Studies in which the rate of radioactive amino acid incorporation into PNMT, followed by precipitation of the enzyme by anti-PNMT antiserum, were performed. These studies revealed that there was no difference between the rate of synthesis of PNMT between the BALB/cJ and the BALB/cN sublines. However, the rate of degradation of PNMT in the BALB/cJ line was half that of the BALB/cN line. Thus it appeared that the reason the BALB/cJ line had twice as many PNMT molecules was that they were being degraded at half the rate as PNMT in the BALB/cN line. It was this altered rate of degradation which was under genetic control and which accounted for the difference in PNMT levels. Similar studies are in progress for TH and DBH.

In recent studies (Barchas, Erdelyi, & Kessler, in preparation), we have investigated strain differences in levels of the enzyme dopamine-β-hydroxylase. The enzyme converts dopamine to norepinephrine. Brains and adrenals of five strains of inbred mice were assayed for dopamine-β-hydroxylase activity by the method of Molinoff, Weinshilboum, and Axelrod (1971). The five strains were: BRT, BALB/cN, SJL, NZB, and CSW.

Brains were dissected into three parts: pons and medulla, brain stem, and cortex. We found no significant strain differences of dopamine-β-hydroxylase activity in pons and medulla, or brain stem. However, dopamine-β-hydroxylase activity in the cortex and adrenals of the five strains varied. Significant differences were found in the cortex dopamine-β-hydroxylase activity between BRT and BALB, BRT and SJL, and NZB and BALB mouse strains. The strain CSW was significantly different from the other strains, with higher enyzme activity in the cortex and lower activity in the adrenal. These results, which will be extended, are very exciting, since one of the major current hypotheses regarding disturbed human behavior involves altered levels of dopamine-β-hydroxylase. Our data suggest potential genetic differences in the regulation of the levels of the enzyme in the brain in animals and, as is true for the other enzymes in the synthetic pathway, requires extension to humans.

In summary, there are wide differences in catecholamine biosynthetic enzyme activity among inbred mouse strains. These differences are genetically determined, and are reflected in the number of molecules of each enzyme. Each enzyme is under the control of a single locus, but phenotypic traits are inherited in association rather than separately, suggesting the existence of a common control gene which regulates their phenotypic expression. In the case of PNMT, a control gene whose product regulates the rate of degradation of the enzyme is proposed.

Strain Differences in Regulatory Mechanisms of Enzymes Involved in the Synthesis of Catecholamines

The work presented demonstrates clear genetically determined differences between strains in the ability to synthesize catecholamines. While static enzyme levels provide limited information, the physiological regulation of enzymatic activity provides a more dynamic means of investigation of control processes. In this regard, we have been particularly interested in the role of PNMT in converting noradrenaline to adrenaline as a model for investigation. As described earlier, while this enzyme is present in high quantities in the mammalian adrenal medulla (Axelrod 1962; Kirshner & Goodall 1957), small amounts have been noted in the brains of rats (Deguchi & Barchas 1971; Barchas et al. 1969; Ciaranello, Barchas, Byers, Stemmle, & Barchas, 1969a; (Pohorecky, Zigmond, Karten, & Wurtman 1969). In rats, PNMT activity in the adrenal has been shown to decline markedly following hypophysectomy; enzyme activity may be restored to normal levels by treatment with ACTH or dexamethasone (Wurtman & Axelrod 1966). Continued steroid treatment (Ciaranello & Black 1971), however, or steroid administration to intact animals (Wurtman & Axelrod 1966) is not effective in elevating the enzyme above normal levels. The half-life of PNMT has been estimated at six days in hypophysectomized rats (Ciaranello & Black 1971). In the intact animals, extraordinary means are generally required to elevate PNMT above normal levels. Elevation of PNMT can be achieved by implantation of an ACTH-secreting tumor (Vernikos-Danellis et al. 1968), unilateral adrenalectomy with compensatory adrenal hypertrophy (Ciaranello et al. 1969b), chronic intermittent immobilization stress (Kvetnansky, Weise, & Kopin 1970), and reflex neuronal stimulation caused by chronic treatment with reserpine (Molinoff et al. 1970) or 6-hydroxydopamine (Mueller et al. 1969). Although reliable measures of PNMT half-life in intact rats are lacking, we have estimated the half-life of the enzyme to be on the order of 20 to 50 days (Vernikos-Danellis et al. 1968).

Because of the lack of rapid changes in PNMT in intact rats, genetic studies which have attempted to focus on glucocorticoid control of the enzyme or stress-induced fluctuations in enzyme activity have not been fruitful. Because of our demonstration that PNMT levels varied widely in the adrenals of several different inbred mouse strains, it was of interest to study these strains further to determine the characteristics of the changes in activity of PNMT in response to stress and to detect strain differences in the physiological regulation of the enzyme.

We first determined whether there were changes in PNMT activity in animals subjected to cold stress by exposure to 4° C for periods up to twelve hours. After three hours of such stress, PNMT levels had increased 1.2-fold and by six hours the levels had increased by 1.4-fold in the C57B1/Ka mouse strain (Ciaranello, Dornbusch, & Barchas 1972c). These findings of a rapid change in the activity of the enzyme which converts norepinephrine to epinephrine as the result of an environmental stress have led to studies to determine whether factors other than environmental stress could alter the activity of the enzyme.

We then wished to compare the rat and the mouse in terms of response to glucocorticoids. Groups of normal DBA/2J mice were injected with dexamethasone, a potent synthetic glucocorticoid. Animals were killed three hours after the administration of this agent, and adrenal PNMT levels determined. The mice treated with dexamethasone showed a striking elevation in enzyme activity after a brief period of exposure to the drug. For comparison, a similar study was performed in intact male Sprague–Dawley rats, 180–200 grams. No change in PNMT activity was detected in the adrenals of these rats (Barchas, Ciaranello, Kessler, & Hamburg, in press).

Since, as mentioned earlier, reflex neuronal stimulation, caused by the administration of reserpine or 6-hydroxydopamine has been shown to cause an elevation in PNMT activity in rats, we decided to test whether mouse adrenal PNMT responded to similar stimulation. Groups of DBA/2J mice were given phenoxybenzamine (10 mg/kg). This alpha-adrenergic blocking agent produces a systemic hypotension which results in the activation of splanchnic nerve inflow to the adrenal medulla. When phenoxybenzamine was given to these mice, a 1.3-fold increase in PNMT activity was seen three hours after administration of the drug.

Altogether the results suggest that in intact mice, PNMT responds relatively rapidly to environmental stress as well as to hormonal and neuronal stimuli. The data suggest that the biologic regulatory controls on PNMT activity may be qualitatively similar in mice and in rats, albeit the rate of response of the enzyme differs profoundly between the two species. Also, in mice, PNMT may play a role in the organism's response to acute stress, whereas in rats, the much slower increase in the enzyme suggests that its biologic role might be involved in the adaptation to chronic stress situations. Lastly, and perhaps the most important point, is that the genetic contribution to the regulation of changes in PNMT activity could be profitably investigated. The rapid nature of the change in mice opens the way for the analysis of genetic differences in catecholamine synthesis using the regulation of PNMT as a model system.

The nature of the physiological regulatory mechanisms of PNMT activity was compared in inbred mouse strains (Ciaranello, Dornbusch, & Barchas 1972d). The strains utilized (DBA/2J, C57Bl/Ka, and CBA/J) showed marked and unexpected differences. All three strains responded to cold stress with an elevation of adrenal PNMT activity, and in all strains the pituitary was involved in the regulation of enzyme activity. However, the mechanism of regulation, in terms of neuronal control of the enzyme, differed among the strains.

In the DBA/2J strain, cold exposure, glucocorticoid administration, and phenoxybenzamine administration were all effective in increasing enzyme activity. The results suggest that in this strain PNMT activity is under both glucocorticoid and neuronal control (Table 3).

In the C57Bl/Ka strain, only cold exposure and adrenocorticotropic hormone (ACTH) were effective in the induction of PNMT. Exogenous glucocorticoid administration had no effect on enzyme activity. In this strain, ACTH appears to exert a direct regulatory effect on the enzyme without the mediation of the adrenal glucocorticoids. No evidence for neuronal control of PNMT activity was found in this strain.

Table 3. Comparison of the Effect of Dexamethasone and Phenoxybenzamine Treatment on PNMT Activities of Two Strains of Mice

Strain	Treatment[a]	Phenylethanolamine N-methyltransferase	p
		unit/adrenal pr[c]	
DBA/2J	Control[b]	0.243 ± 0.016	
	3-hr dexamethasone	0.320 ± 0.014	< 0.01
C57B1/Ka	Control	0.143 ± 0.101	
	3-hr dexamethasone	0.149 ± 0.009	not significant
DBA/2J	Control	0.173 ± 0.011	
	phenoxybenzamine	0.217 ± 0.014	< 0.02
C57B1/Ka	Control	0.125 ± 0.003	
	phenoxybenzamine	0.110 ± 0.006	not significant

[a] Dexamethasone (5 mg/kg) or phenoxybenzamine (10 mg/kg) was administered intraperitoneally to groups of mice.
[b] Untreated controls (not shown here) were also studied; they were no different from the NaCl-treated animals.
[c] Enzyme activity units are the means and standard errors of 8–10 animals/group.

In the CBA/J strain, cold exposure increased the level of the enzyme. The enzyme responded to ACTH, but the response was mediated by glucocorticoids. There is no evidence of neuronal control of the enzyme.

The half-life of the enzyme was estimated to be one hour in the DBA strain, three hours in the C57B1/Ka strain, and seven hours in the CBA/J strain. The rate of increase of enzyme activity following induction is ten times greater in the DBA/2J strain than in either the C57B1/Ka or CBA/J strains. These findings suggest that genetic differences in the breakdown and/or the utilization of PNMT may also be present. These various differences are shown in Table 4.

Approaches to the Psychological and Biochemical Study of Depression Which Include the Genetics of Neuroregulatory Systems

In the material that follows we would like to focus on aspects of behavioral biology that relate to anger and depression and the relations of those emotions to catecholamines and the regulation of catecholamines. We will first focus on psychological aspects in primates and humans and then proceed to the physiological studies. (We have considered these areas in greater detail elsewhere; Hamburg, Hamburg, & Barchas, in press.)

The adaptive function of primate groups alerts us to look for processes in the individual that facilitate the development of inter-individual bonds. Such bonds as status orders and affiliative bonds are rapidly established in newly formed groups of nonhuman primates (Barchas 1971).

Table 4. Comparison of Factors Which Increase PNMT Activity in Different Mouse Strains

	Cold exposure	Glucocorticoid	Neuronal	Direct ACTH
DBA/2J	+	+	+	0
C57Bl/Ka	+	0	0	+
CBA/J	+	+	0	0

Key: 0 = no increase
       + = increase

We find useful guidance in the concept that individuals seek and find gratifying those situations that have been highly advantageous in survival of the species. Tasks that must be done for species' survival tend to be quite pleasurable; they are easy to learn and hard to extinguish. Their blockage or deprivation leads to anger, coping behavior, and (if prolonged) depression. Such blockage is often accompanied by physiological responses that support actions necessary to correct the situation. In the adult human, a remarkable variety of coping behavior may be mobilized by such blockage or deprivation, determined in substantial part by cultural patterning (Hamburg & Adams 1967). In view of the extreme dependence on learning in the human species, such bonds would most likely be greatly strengthened through learning. Selection may operate on differential readiness for learning responsiveness and attachment to others of the same species.

Another evolutionary aspect of emotion is that individuals avoid and find distressing those situations that have been highly disadvantageous in species' survival. Applied to the specific issue of inter-individual bonds, it seems reasonably clear that disruption of such bonds in primates is perceived as seriously threatening. It is usually felt as unpleasant and is often associated with physiological responses of alarm and mobilization. Such disruptive events usually stimulate coping behavior that tends toward restoration of strong bonds.

In the case of emotions experienced as distressing, the associated actions have also been linked with survival, phylogenetically or ontogenetically or both. Once again, the message is "This is important," but the further message is different: "It is important in a bad and undesirable way, it should be avoided in the future." Thus, distressing experiences may be quite useful though unpleasant. They have a signal function which warns the organism and other individuals significant to him that something is wrong, attention must be paid, learning capacities utilized, resources mobilized to correct the situation.

Let us briefly consider anger and depression from this perspective. The angry organism is making an appraisal of his current situation, which indicates that his immediate or long-run survival needs are jeopardized, his basic interests are threatened. Moreover, his appraisal indicates that another organism (or group) is responsible for this threat. While there are many ways he can go from this appraisal, the general tendency is to prepare for vigorous action to correct the situation, quite likely action directed against the person(s) seen as causing or at least manifesting the jeopardy to his needs. The signals are likely to be transmitted to these individuals as well as to the person's own decision-making apparatus. The significant others are then likely to respond in a way that will ameliorate the situation. In a medium range of intensity, anger and some associated aggressive actions are likely to bring about a result desirable to the person and acceptable to his significant others. At very high intensity, the risk of serious injury becomes great for the initiator as well as for others. This behavior can readily become maladaptive.

Depressive responses have similar characteristics. However, they tend to follow a prior angry period; but the angry responses have not elicited a rewarding outcome. Then a feeling of sadness and discouragement sets in. The subject estimates the probability of effective action as low. By the term "effective action" we refer here to action the subject believes to be in his self-interest or group interest, even though his belief may be vaguely formulated. He may, in effect, have been prepared for this orientation through the long past experience of his species or his population or his family life or his own experience—or some combination of these. But however he came to this appraisal, it is now a firm commitment, somehow bound up with his survival. How can the depressive responses be viewed as adaptive? As we saw in the case of anger, they can be adaptive in a medium range of intensity. His feelings of sadness and discouragement may be a useful stimulus to consider ways of changing his situation. If a key human relationship is in jeopardy, ways of improving that relationship, or substituting a better one, may be considered. Moreover, his state of sadness may elicit heightened interest and sympathetic consideration on the part of significant other people. Their actions as well as his own may work toward improvement of the situation. But at very high intensity, the depressive responses increase survival risks for the person: (a) in terms of his own behavior, physiology, and susceptibility to disease; (b) in terms of the responses of others, which tend to become unfavorable or at least ineffective in the face of intense depression (Klerman 1971).

So for both anger and depression there appears to be a curvilinear relation between intensity of the emotional response and effectiveness of behavior in adaptation. At low and moderate intensities survival probabilities are increased, but at very high intensities they are lowered. This concept is offered as a statistical proposition covering many occasions of anger and depression on the part of many individuals in many groups.

In several ways, observations of reaction to inter-animal loss in the natural habitat are consistent with recent laboratory investigations of a separation-induced model of depressive behavior in monkeys (Seay, Hansen, & Harlow 1962; Seay & Harlow 1965; Spencer-Booth & Hinde 1967; Hinde & Spencer-Booth 1968; Kaufman & Rosenblum 1967; Rosenblum & Kaufman 1968; Jensen & Tolman 1962; Mitchell, Harlow, Griffin, & Moller 1967; Van Lawick-Goodall 1968, 1973). In these experiments a mother and infant are separated. Thus a strong bond is disrupted. Though there are variations, the main tendency is toward a depression-like response in both mother and infant. In the infants, typically there is initial distress: calling and searching for a day or two. This is followed by greatly diminished activity, decreased play, huddling posture, and decreased food intake. They resemble humans who report feeling depressed. There are some indications of lasting effects due to brief separation.

Is there anything comparable to this depressive syndrome induced by separation or loss in humans? Over many years, some clinicians have made sensitive observations of personal loss, grief, and clinical depression (Lindemann 1944; Bowlby 1969, 1973). Patients often come to medical attention in the context of an important loss. The most vivid circumstance is the grief reaction to the loss of a personally significant individual (McConville, Boag, & Purohit 1972). The grief reaction is a specific pattern of distress in which the person's focus is on the loss. Usually, gradual recovery occurs through a difficult process of mourning over a period of months. However, some persons slide into a clinical depression in which there is a pervasive undermining of prior interests and human relationships, with feelings of dependency.

In the past few years, several research groups have undertaken systematic study of possible relations between life events and the onset of depressive episodes serious enough to come to psychiatric attention. Some of these studies have utilized appropriate control or comparison groups (Klerman 1971; Clayton, Desmarais, & Winokur 1968; Clayton 1973; Beck & Worthen 1972; Cadoret, Winokur, Dorzab, & Baker 1972; Clancy, Crowe, Winokur, & Morrison 1973; Leff, Roatch, & Bunney 1973; Paykel, Myers, Dienelt, Klerman, Lindenthal, & Pepper 1969; Parkes 1972).

Thus, the present evidence indicates that: (1) the experience of interpersonal loss elicits much distress—grief is one of the most difficult transitions of the life cycle; (2) it is a common precipitating factor in clinical depressions; (3) but such depressions need not be triggered by separation or loss; (4) separation or loss may trigger distress reactions other than depression (Holmes & Rahen 1967); (5) most occurrences of separation and loss do not precipitate clinical depression—the best available evidence suggests that only 10 to 20 percent of individuals confronted with separation or loss of major personal significance develop a clinical syndrome of any kind.

A few studies have explored ways of coping with separation and loss (Hamburg, Adams, & Brodie, in press). The variety and effectiveness of such patterns in the general population is impressive. These studies highlight the question of differential susceptibility to loss and grief. What is the special vulnerability of those who become overwhelmed with despondency?

In monkeys, apes, and humans there are marked individual differences in the response to separation or loss. Some get more depressed than others, and these variations deserve biological scrutiny. Do those who get very depressed differ in some way in respect to brain amines or other putative transmitters or endocrine function from those who were much less susceptible to the depression-like response? Both the primate separation experiments and clinical observations point the way toward a model of depression in which behavioral and biological linkages can be investigated. They highlight the question of differential susceptibility to experiences of separation and loss. Since some individuals are more vulnerable to such stresses than others, what biological factors might be relevant to these individual differences?

The many determinants of individual differences in stress response cannot be discussed here. It is an enormous and fascinating area, much of which remains to be explored. Some people are inclined to react with anger, others with depression. The same person may respond angrily to stress on one occasion, depressively on another. We know that these and other emotional responses are mediated to a great extent by circuits of the limbic-hypothalamic-midbrain region of the brain (MacLean 1970). These circuits are influenced significantly by hormones; and since transmission of impulses in these circuits depends heavily on biogenic amines in this brain region, we shall turn our attention to that area. Before doing so, we wish to note the important influences of the individual's prior experience on subsequent inclinations to anger and depression. We have in some respects discussed these elsewhere (Hamburg & Hamburg, in press; Hamburg 1969; Hamburg & Van Lawick-Goodall, in press). The fact that we do not pursue these influences here should in no way diminish the great importance we attach to them.

The biogenic amines have been shown to be particularly important in stress states. Both the catecholamines and the indoleamines are found in very discrete areas of the brain thought to be involved in emotional behaviors and are affected by many drugs known to alter behavior. This approach has received further support from investigation of changes in the concentration or utilization of these neuroregulators in the brain as a consequence of behavioral states.

Stressful experimental procedures cause significant alterations in brain biogenic amine levels (Barchas & Freedman 1963). Among these procedures are forced swimming, cold exposure, electric foot shock, and immobilization. Brain biogenic amine levels would be expected to change if (1) formation (anabolism) of the amines was increased or decreased; (2) destruction (catabolism) of the amines was enhanced or suppressed; or (3) a combination of the two processes occurred. The development of the procedures for measuring dynamic changes in brain biogenic amine metabolism revealed that most of the above stressors affecting endogenous amine levels did so by increasing the turnover and metabolism of norepinephrine, dopamine, and serotonin. Interestingly, very few instances of a stress-induced decrease in catecholamine or serotonin turnover are known, these being due to hypothermia, after certain forms of shock stress in rats and social isolation in mice.

Technical difficulties have so far limited our knowledge of the effects of stress on brain amines in humans. Thus, for example, we do not know whether various types of stress alter the rate of synthesis or utilization of brain amines and whether there are times when synthesis cannot keep up with the need for these compounds. Clearly, such an issue is of great importance, since there are many forms of psychiatric disorders in which an inability to respond to stress is a prominent difficulty. Genetically determined biochemical abnormalities in brain amine metabolism may well turn out to have clinical significance; we shall return to this point later.

There are several ways in which amine neuroregulatory agents in the central nervous system can be related to emotional behavior with particular application to human studies. The most important has been the measurement of metabolites of the neuroregulators in the spinal fluid, although the principal neuroregulators themselves are not present in spinal fluid. Further examination of such material and the dynamic processes involving them are particularly important—especially in relation to changes over time that may occur concomitantly with behavioral changes. For example, the same person may be studied when he is depressed and not depressed. The development of mass fragmentography and the use of staple isotopes should prove a particularly powerful tool for such studies (DoAmaral 1973).

There is much current interest in the relationship of catecholamines to depression, especially in the catecholamine hypothesis of affective disorders. Briefly, the hypothesis relates depression to a relative deficiency of norepinephrine at certain central synapses, and relates the manic states to a relative excess of norepinephrine (Kety & Schildkraut 1967). The hypothesis is consistent with much of the available evidence and has been a powerful stimulus to both basic research on brain biochemistry and clinical research or psychopharmacology. Nevertheless, some of the evidence also supports other notions as well, including the opposite view of a relative excess of catecholamines in relation to depression (Wyatt, Portnoy, Kupfer, Snyder, & Engelman 1971).

Some investigators have pointed out that much of the evidence for the catecholamine hypothesis is indirect and could implicate other neuroregulatory agents as well. There is a competing indoleamine hypothesis which implicates serotonin for which recent studies have been drawn together in a volume edited by Barchas and Usdin (1973). While is is exceedingly difficult to sort out the evidence, many investigators favor the view that serotonin systems in the brain play a crucial role in regulating responsivity to environmental stimuli. Procedures that lower brain serotonin tend to produce hyperresponsivity to both internal and external stimuli (Conner, Stolk, Barchas, & Levine 1970). This occurs in a variety of species, including man. Such hyperresponsivity may be manifested in disturbances of sleep, increased susceptibility to convulsions, greater irritability and aggressiveness (Lipton 1973). The picture of serotonin functions in the brain is so complex that resolution must await further research, but it may well be that balances or interactions between catecholamine-mediated neurons and indoleamine-mediated neurons will prove to be important in the regulation of mood and the emotional experiences of anger and depression.

Changes have been reported in catecholamine patterns in relation to depression which are of considerable interest. Several investigators have published data indicating that during periods of depression certain patients excrete significantly less of a norepinephrine metabolite, MHPG, than when they are not feeling depressed. Moreover, those patients who excrete less MHPG can be identified by their emotional responses to tricyclic antidepressants or amphetamine: they become less depressed. Maas, who has been particularly interested in this area, has recently reviewed the topic and reported new findings (Maas, Dekirmenjian, & Jones 1973).

Schildkraut's group has done much work on the catecholamine hypothesis and has reported that urinary excretion of MHPG is significantly lower in patients with bipolar depressions than in patients with chronic characterological depressions. Excretion of MHPG was not related to the degree of retardation, agitation, or anxiety in these patients (Schildkraut, Keeler, Papousek, & Hartman 1973).

These studies are consistent with the concept that a subgroup of depressed patients may be identified clinically (primary affective disorders), biochemically (low MHPG), and pharmacologically (depression-relieving response to desipramine, imipramine, and amphetamine). It is at least plausible that these persons have an abnormality in metabolism and/or disposition of norepinephrine which predisposes them to intense depression.

With regard to anger and aggressive behavior, extensive studies in animals and humans have provided strong links to brain biogenic amines. Amphetamines, which are thought to act through catecholamine mechanisms (Snyder 1972, 1973), commonly elicit paranoid reactions (Griffith, Cavanaugh, & Oates 1970; Angrist & Gershon 1970; Tinklenberg 1971). Although responses to amphetamine are complex, one major component is intense anger. In this respect, the syndrome differs considerably from those produced by other psychotomimetic drugs.

Studies in animals utilizing a variety of test situations have suggested a very powerful role of catecholamines in aggressive behavior. Catecholamines are essential for many forms of aggressive behavior and catecholamine mechanisms are markedly altered by aggressive behavior (Reis, in press;

Reis & Fuxe 1969; Reis & Gunne 1965; Welch & Welch 1969*a,b,c*; Eichelman & Thoa 1973; Eichelman, Thoa, Bugbee, & Ng 1972; Stolk, Conner, Levine, & Barchas 1974). It is of interest that there appears to be an association between an increased level of fighting behavior and increased levels of tyrosine hydroxylase, when levels of the enzyme, as reported earlier in this paper, are compared to levels of fighting behavior in the different strains of mice.

Thus, to the extent that processes involving anger and depression involve catecholamines, genetic regulation of these compounds becomes of great importance. As in the case of the behavior-endocrine-genetic approach to stress responses that we suggested some years ago, it seems desirable to view the amine systems dynamically, responding over time to loads placed upon the organism. Genetic factors in amine metabolism may make it more difficult for some persons to sustain an effective response to prolonged stress, more likely to explode in anger or slide into despondency.

Genetically oriented research is helping to sort out distinctive subgroups of depressive disorders (Angst & Perris 1972; Winokur, Morrison, Clancy, & Crowe 1973). Other important studies have come from the group of Fieve (Mendlewicz, Fieve, Rainer, & Fleiss 1972*a*; Mendlewicz, Fieve, Stallone, & Fleiss 1972*b*; Mendlewicz, Fleiss, & Fieve 1972*c*; Fieve, this volume). Experience in other fields of clinical investigation suggests that reliable differentiation of subgroups can be helpful in clarifying different underlying mechanisms.

Genetically determined variation involving biogenic amine mechanisms and their physiological regulation may provide a basis for significant individual differences in emotional and endocrine responses to stressful situations. Individuals with certain gene variants may, in the face of severe life stresses, be predisposed to depressive reactions or periods of labile emotion, including episodes of intense anger. We consider it likely that some severe emotional disorders involving intense anger or depression may be partly based upon genetically determined alterations in normal biochemical processes. These biochemical predispositions must interact in complex ways with environmental factors such as separation, loss, or other jeopardy to crucial human relationships.

The catecholamine hypothesis of depression provides a sense of the need for further knowledge in the new field of behavioral neurochemistry. Regardless of whether one accepts a relative excess or the relative deficiency of the neuroregulatory agent, we will have to go further in our recognition that the same compound in different areas of brain may have markedly different and even contradictory roles. The notion of a relative deficiency or excess has proven to be a valuable one. Hopefully, by the end of this decade, we will be able to evaluate these changes vis-à-vis our understanding of transmitter formation, coupling of nerve excitation to release processes, rate of transmitter release and reuptake, synaptic receptors, and the biochemical effects of receptor excitation. And, instead of concentrating on only one neuroregulatory agent, one might more appropriately investigate the interactions between neuroregulatory agents in important brain regions.

A diagram of the model catecholamine neuron is shown in Figure 3. We have examined only the first step, the rate-limiting enzyme, tyrosine hydroxylase, and found powerful genetic controls. We expect that for each of the steps involved in catecholamine mechanisms there are genetic controls that would alter the activities of the processes or their ability to respond to environmental processes. Thus, as only one example, one could expect that such genetic controls would determine the extent to which an enzyme such as tyrosine hydroxylase could be activated in times of stress or the degree to which new synthesis of the enzyme could occur in times of prolonged stress. In the simple model of relative deficiency of catecholamines, one could imagine that stress-induced utilization of catecholamines causes or requires biochemical changes which later produce in susceptible persons the longer-term behavior which we recognize as depression.

Fig. 3. Functional Model of a Noradrenergic Neuron.

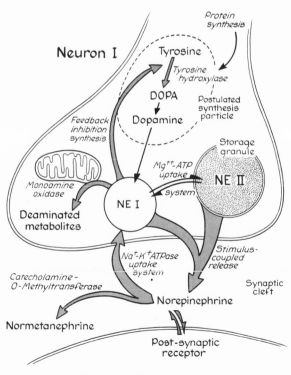

Models for such behaviors can include increases or decreases of relative neuronal activity. One could imagine that in a situation involving a prolonged emotional state, an inability of the enzyme to respond would produce a relative deficiency of the transmitter. On the other hand, perhaps a long-term state would produce an excess of the enzymatic machinery, which, if it were not regulated by another enzyme capable of destroying the synthesizing enzyme (Ciaranello & Axelrod 1973), might cause a relative excess of transmitter which could produce emotional behaviors inappropriate to those in which the individuals later found themselves. Again, very speculatively, such a relative excess might be produced in a prolonged state of anger and be continued when the "cause" of the anger had passed. The emotional handling of the continued biochemical message of anger would be coped with by directing the anger toward the self, as well as outwardly, with resultant symptoms of psychological depression. Many other models involving aspects of catecholamine processes can be constructed for a variety of normal and abnormal behaviors with interrelation of developmental processes. For example, genetic variation in mechanisms involving receptors or reuptake, or alterability of those mechanisms with behavioral states, provides another case in which one could postulate changes involving either increases or decreases with environmental effects and long-term behavioral processes. Taken as a whole, it is apparent that in the investigation of both biochemical and behavioral parameters, and the manner in which they interact, genetic factors which are involved in regulatory mechanisms will become extremely important.

Toward a Behavioral Neurochemistry

Recent research, including some presented in this meeting, suggests the development of a new field of behavioral neurochemistry. Such a term is analogous to physiological psychology and yet recognizes the new approaches and techniques which will be necessary for the field. The area is exciting in its conception, posing many problems from the technical to the philosophical. Some of the problems include issues of delineation of the action of drugs, relation of biochemical events to psychological events, lack of basic information about the relative importance of various metabolic pathways, the determination of mechanisms of regulation, the importance of genetic processes, and the application of new knowledge to human disorders.

The regulation of important neuroregulatory agents and how such regulation can relate to behavior has now become the most important set of problems in the field at a basic level. Thus, for several decades much biochemical research related to psychiatry was concerned with finding the "abnormal metabolite" related to behavior; in the future one might expect to find more emphasis on the regulation of normal processes which relate to behavior and, at some times in some individuals, may result in alterations which could change behavior. For example, many theories of abnormal behavior have postulated methylated indole derivatives, but we know very little about the regulation of many of the enzymatic processes, such as methylation (Ciaranello et al. 1972, Deguchi & Barchas 1971), to mention just one step.

The area of genetic studies is one which may be particularly important to behavioral investigations. To date there has been limited investigation of whether there are differences in enzyme activities, metabolic pathways, utilization rates, or responses to stress using different strains of inbred species. Studies have shown that different genetic strains have a considerable variation of the steroid hormones produced by the adrenal cortex (Hamburg 1967; Hamburg & Kessler 1967; Hamburg 1970) and thyroid hormones (Hamburg & Lunde 1967). A number of illnesses have been demonstrated in which there is a genetic difference in formation of adrenal cortical hormones; several of these illnesses lead to marked behavorial changes.

An example of a potential interaction between genes, biogenic amines, and behavior is contained in the findings of Schlessinger et al. (1965). The investigators demonstrated that inbred mouse strains susceptible to audiogenic seizures had lower levels of brain serotonin and norepinephrine at twenty-one days of age (when they are most susceptible to seizures).

The catecholamines formed both in the periphery and in the brain have been strongly linked to behavioral states (Glowinski & Baldessarini 1966; Barchas et al. 1971; Schildkraut & Kety 1967; Bunney, Brodie, Murphy, & Goodwin 1971; Hamburg 1970; Barchas et al. 1972a). Many questions still remain to be investigated regarding the association between these neuroregulatory agents and long-term behavioral changes. To date there have been limited studies including dynamic processes, such as the effects of various stressors on the synthesis, turnover, or utilization of catecholamines in relation to genetic factors. Ultimately, these questions have to be related to man in terms of the role of genetic factors in the regulation of the response to stress. Alterability of the processes involved in catecholamine formation, release, and in metabolic pathways (both short-term and long-term) is essential if these compounds are indeed involved in emotional behavior.

An extensive listing could be made of how possible genetic differences in catecholamine production, utilization, etc., both in the adrenal or in the brain, might affect behavior. It has been shown that in response to stress, differing amounts of adrenal cortical steroids are released in different genetic strains. It is easy to imagine that in response to stress, two individuals might send the same number of nerve impulses to the adrenal medulla, but because of a genetic difference one individual might form differential amounts of catechols, release differential amounts, metabolize the catechols at different rates, or have differential passage across the blood-brain barrier. If any of these possibilities were to occur, clearly, there could be behavioral changes when the adrenaline reached various target organs, including the brain.

Other possibilities to be investigated include the presence of minor or abnormal pathways in adrenal catechol metabolism, including the physiological controls on those pathways and how they might be altered by stress. Analogous factors could be relevant not only to adrenal catecholamines as hormones related to behavior but also to brain catecholamines and other putative transmitters between nerve cells.

Examples of the overwhelming need for additional knowledge regarding genetic aspects of regulation of processes involving putative neurotransmitters can be seen in the need to extend the current studies. It will be necessary to determine the extent to which the differences noted in the physiological regulation of epinephrine formation are due to genetic factors and the nature of the genetic control over these mechanisms. In addition, both for the levels of the enzymes in unstressed and in stressed animals it will be important to determine the extent to which the enzymatic changes are associated with differences in the rate of secretion of epinephrine and the long-term response to a variety of stressors. Also, the information already obtained opens the way to learning more about the genetic linkage of the various enzymes involved in catecholamine formation and the location of the relevant genes.

Other directions in which the work might be extended include the study of the genetic factors involved in developmental processes in relation to the catecholamines (Milkovic, Deguchi, Winget, Barchas, Levine, & Ciaranello, 1974). It will be important to study the effects of early stress on adrenal medullary function and to determine whether any of the regulatory mechanisms are modified by early experience. While we have studied the regulation of one of the key enzymes, PNMT, studies of the regulation of TH and other enzymes involved in the formation of catecholamines need to be undertaken. Eventually, means will have to be found to evaluate brain catecholamine function and adrenal medullary function in humans in terms of differential enzyme activities and regulatory mechanisms in relation to genetic factors. Such steps would include the study of electrophoretic variants of the various enzymes, their ontogenesis, and their function under contingencies of stress.

It is difficult at this early point in the development of psychoendocrine and biochemical-genetic research to give an adequate general formulation of the potential relations between biochemistry and behavior. At this stage, it seems reasonable to assume that biochemical and psychological processes are intimately related. Biochemical processes may affect, for example, activity levels, emotional tones, and the susceptibility to stress. Psychological processes may influence biochemical processes, for example, by producing shifts between pathways, altering utilization of neuroregulators, and inducing enzymatic changes. Such biochemical events might lead to further psychological changes or to the maintenance of psychological states. While we do not assume that emotional distress is primarily caused by biochemical abnormalities, it is also true that biochemical events may profoundly alter the ability of the organism to respond to its environment.

On the one hand, one encounters those who view mental illness and mind as apart from the brain, and, on the other, those who feel that "a twisted thought, a twisted molecule." While granting that there are clearly biochemical and biophysical processes involved in all mental activity, those processes may have no deviation from normal in thought processes ranging from learning to philosophical viewpoint. Yet, biochemical processes may involve later behavior and emotional processes by setting an emotostat in one or another way, and behavior may change biochemical emotostats (Barchas 1973). What is needed is a view which recognizes the subtle interplay of psychological, biochemical, and physiological processes.

It is clear that more information relative to the issues of the manner in which psychological events interact with biochemical variables will be needed. How do developmental patterns and behavioral states influence biochemistry? Can the response of neuroregulatory agents to stress later in life be altered by early experience? What types of changes in the various mechanisms involving neuroregulatory agents can be found in different behavioral states? What are the short- and long-term biochemical effects of different types of behavioral states? What is the possibility that behaviors, such as repeated aggressive behavior, may alter the propensity for long-term or repeated episodes of the behavior? Such studies involve animal investigation, but may lead to human studies. For example, to what extent does a psychological event, e.g., anxiety, trigger biochemical changes which might affect later psychological events thereby causing a cycle of actions which is neither wholly "psychological" nor wholly "biochemical"?

One might well imagine that there will be illnesses which will meet the model of phenylketonuria in which a set of clear-cut definable biochemical changes lead to severe behavioral changes. Understanding why these behavioral changes occur will be a crucial step. On the other hand, there may be other emotional illnesses which are more purely psychological than in the model just mentioned and there may be still other illnesses in which a set of psychological events in an individual with appropriate biochemical structures leads to an illness.

If one visualizes a relationship between biochemical and psychological events and assumes changes in response to a psychological event, then such changes become very important in relation to the length of time that they persist and the manner in which environmental and genetic factors interrelate. To account for long-term emotional behaviors and alterability of behaviors, one would have to assume alterability in the processes under-lying chemical events. In the simplest model, we would assume that psy-chological events affect the body chemistry and that the chemical change in turn affects the future psychological events. Thus, with particular genetic predispositions, chemical changes that are long-term in nature could, in effect, "lock in" certain psychological sets.

Genetic influences on the physiological regulation of catecholamines might be involved in the differential response to stress and differential coping mechanisms in behaviors such as grief and loss, thus influencing the likelihood of depression.

The presence of genetically determined variation involving the biogenic amines and their physiological regulation may provide a basis for the in-dividual differences of endocrine and behavioral responses to stressful situations. Individuals with gene variants may, in the face of chronic life stresses, be predisposed to depressive episodes or periods of labile affect. We consider it likely that some severe emotional disorders, including some forms of manic-depressive illness, depression, schizophrenia, and autism, may be based upon genetically determined alterations in normal biochemical processes. These biochemical predispositions may interact in complex ways with environmental factors, such as severe psychological stresses.

The present studies, taken together, represent the first demonstration that the activities of the enzymes involved in catecholamine biosynthesis are subject to marked variation as a consequence of genetic differences. The experiments also provide the first suggestion that major differences in the physiological regulation of an enzyme involved in the formation of catecholamines exist within a single species. Such findings point out that conceptualization in terms of unitary regulatory mechanisms is illusory; within a single species, multiple mechanisms may be operative. Thus, our findings reveal a store of considerable physiological variation in the activity and regulatory control of the enzymes involved in catecholamine formation, which may have important implications for understanding the broad spec-trum of affective and psychological response of organisms to stress and the ultimate effects of stress on the physiological and behavioral functioning of the organism.

References

Alousi, A., & Weiner, N. The regulation of norepinephrine synthesis in sympathetic nerves: effect of nerve stimulation, cocaine and catecholamine-releasing agents. *Proceedings of the National Academy of Sciences (Washington, D.C.),* 1966, **56**, 1491–1496.

Angrist, B. M., & Gershon, S. The phenomenology of experimentally induced amphetamine psychosis—preliminary observations. *Biological Psychiatry,* 1970, **2**, 75–107.

Angst, J., & Perris, C. The nosology of endogenous depression. Comparison of the results of two studies. *International Journal of Mental Health*, 1972, **1**, 145–158.

Axelrod, J. Purification and properties of phenylethanolamine N-methyl-transferase. *Journal of Biological Chemistry*, 1962, **237**, 1657–1660.

Axelrod, J., & Vesell, E. S. Heterogeneity of N- and O-methyltransferases. *Molecular Pharmacology*, 1970, **6**, 78–84.

Badr, F. M., & Spickett, S. G. Genetic variation in the biosynthesis of corticosteroids in *Mus musculus*. *Nature*, 1965, **205**, 1088–1090.

Barchas, J., Ciaranello, R. D., Kessler, S., & Hamburg, D. A. Genetic aspects of the synthesis of catecholamines in the adrenal medulla. In R. P. Michael (Ed.), *Hormones, the brain and behaviour*. London: S. Karger, in press.

Barchas, J. D., Ciaranello, R. D., & Steinman, A. M. Epinephrine formation and metabolism in mammalian brain. (A. E. Bennett Award Paper, 1968.) *Biological Psychiatry*, 1969, **1**, 31–48.

Barchas, J. D., Ciaranello, R. D., Stolk, J. M., Brodie, H. K. H., & Hamburg, D. A. Biogenic amines and behavior. In S. Levine (Ed.), *Hormones and behavior*. New York: Academic Press, 1972, pp. 235–329. (a)

Barchas, J. D., Erdelyi, E., & Angwin, P. Simultaneous determination of indole- and catecholamines in tissues using a weak cation exchange resin. *Analytical Biochemistry*, 1972, **50**, 1–17. (b)

Barchas, J. D., Erdelyi, E., & Kessler, S. Strain differences in dopamine-β-hydroxylase activity in mice. In preparation.

Barchas, J. D., & Freedman, D. X. Brain amines: response to physiological stress. *Biochemical Pharmacology*, 1963, **12**, 1232–1235.

Barchas, J. D., Stolk, J. M., Ciaranello, R. D., & Hamburg, D. A. Neuroregulatory agents and psychological assessment. In P. McReynolds (Ed.), *Advances in psychological assessment*. Vol. 2. Palo Alto, California: Science and Behavior Books, 1971, pp. 260–292.

Barchas, J., & Usdin, E. (Eds.), *Serotonin and behavior*. New York: Academic Press, 1973.

Barchas, P. R. Differentiation and stability of dominance and deference orders in rhesus monkeys. Unpublished doctoral thesis, Stanford University, 1971.

Barchas, P. R. Approaches to aggression as a social behavior. In R. Ofshe (Ed.), *Interpersonal behavior in small groups*. Englewood Cliffs, N.J.: Prentice Hall, 1973, pp. 388–401.

Beck, J. D., & Worthen, K. Precipitating stress, crisis theory, and hospitalization in schizophrenia and depression. *Archives of General Psychiatry*, 1972, **26**, 123–129.

Bowlby, J. *Attachment and loss*. Vol. 1. *Attachment*. London: Hogarth, 1969.

Bowlby, J. *Attachment and loss*. Vol. 2. *Separation: Anxiety and Anger*. London: Hogarth Press and the Institute of Psycho-Analysis, 1973.

Bunney, W. E., Brodie, H. K. H., Murphy, D., & Goodwin, F. Studies of α-CH_3-p-tyrosine, L-DOPA, and L-tryptophan in depression and mania. *American Journal of Psychiatry*, 1971, **127**, 872–881.

Cadoret, R. J., Winokur, G., Dorzab, J., & Baker, M. Depressive disease: life events and onset of illness. *Archives of General Psychiatry*, 1972, **26**, 133–136.

Chai, C. K., & Dickie, M. M. Endocrine variations. In E. L. Green (Ed.), *Biology of the laboratory mouse*. New York: McGraw–Hill, 1966, pp. 387–403.

Ciaranello, R. D. Regulation of phenylethanolamine N-methyltransferase. In S. H. Snyder & E. Usdin (Eds.), *Frontiers in catecholamine research.* New York: Pergamon, 1973, pp. 101–105.

Ciaranello, R. D., & Axelrod, J. Genetically controlled alterations in the rate of degradation of phenylethanolamine N-methyltransferase. *Journal of Biological Chemistry*, 1973, **248**, 5616–5623.

Ciaranello, R. D., Barchas, R. E., Byers, G. S., Stemmle, D. W., & Barchas, J. D. Enzymatic synthesis of adrenaline in mammalian brain. *Nature*, 1969, **221**, 368–369. (a)

Ciaranello, R. D., Barchas, R., Kessler, S., & Barchas, J. D. Catecholamines: strain differences in biosynthetic enzyme activity in mice. *Life Sciences (Part I)*, 1972, **11**, 565–572. (a).

Ciaranello, R. D., Barchas, J. D., & Vernikos-Danellis, J. Compensatory hypertrophy and phenylethanolamine N-methyl transferase (PNMT) activity in the rat adrenal. *Life Sciences (Part I)*, 1969, **8**, 401–407. (b)

Ciaranello, R. D., & Black, I. B. Kinetics of the glucocorticoid-mediated induction of phenylethanolamine N-methyl transferase in the hypophysectomized rat. *Biochemical Pharmacology*, 1971, **20**, 3529–3532.

Ciaranello, R. D., Dankers, H. J., & Barchas, J. D. The enzymatic formation of methanol from S-adenosylmethionine by various tissues of the rat. *Molecular Pharmacology*, 1972, **8**, 311–317. (b)

Ciaranello, R. D., Dornbusch, J. N., & Barchas, J. D. Rapid increase of phenyl-ethanolamine-N-methyltransferase by environmental stress in an inbred mouse strain. *Science*, 1972, **175**, 789–790. (c)

Ciaranello, R. D., Dornbusch, J. N., & Barchas, J. D. Regulation of adrenal phenylethanolamine N-methyltransferase activity in three inbred mouse strains. *Molecular Pharmacology*, 1972, **8**, 511–520. (d)

Ciaranello, R. D., Hoffman, H. J., Shire, J. G. M., & Axelrod, J. Genetic regulation of the catecholamine biosynthetic enzymes II. Inheritance of tyrosine hydroxylase, dopamine-β-hydroxylase, and phenylethanolamine-N-methyltransferase. *Journal of Biological Chemistry*, 1974, **249**, 4528–4534.

Clancy, J., Crowe, R., Winokur, G., & Morrison, J. The Iowa 500: precipitating factors in schizophrenia and primary affective disorder. *Comprehensive Psychiatry*, 1973, **14**, 197–202.

Clayton, P. J. The clinical morbidity of the first year of bereavement: a review. *Comprehensive Psychiatry*, 1973, **14**, 151–157.

Clayton, P., Desmarais, L., & Winokur, G. A study of normal bereavement. *American Journal of Psychiatry*, 1968, **125**, 168–178.

Conner, R. L., Stolk, J. M., Barchas, J. D., & Levine, S. *Para*-chlorophenylalanine and habituation to repetitive auditory startle stimuli in rats. *Physiology and Behavior*, 1970, **5**, 1215–1219.

Coupland, R. E. On the morphology and adrenaline-noradrenaline content of chromaffin tissue. *Journal of Endocrinology*, 1953, **9**, 194–203.

DoAmaral, J. R. An approach to the assay of serotonin using gas-liquid chromatography and mass fragmentography. In J. Barchas & E. Usdin (Eds.), *Serotonin and behavior.* New York: Academic Press, 1973, pp. 201–207.

Deguchi, T., & Barchas, J. Inhibition of transmethylations of biogenic amine by S-adenosylhomocysteine. *Journal of Biological Chemistry*, 1971, **246**, 3175–3181.

Eichelman, B., & Thoa, N. B. The aggressive monoamines. *Biological Psychiatry*, 1973, **6**, 143–164.

Eichelman, B., Thoa, N. B., Bugbee, N. M., & Ng, K. Y. Brain amine and adrenal enzyme levels in aggressive, bulbectomized rats. *Physiology and Behavior*, 1972, **9**, 483–485.

Euler, U.S. v. Problems in neurotransmission. In C. F. Cori, V. G. Folgia, L. F. Leloir & S. Ochoa (Eds.), *Perspectives in biology*. New York: Elsevier, 1962, pp. 387–394.

Friedman, S., & Kaufman, S. 3,4-Dihydroxyphenylethylamine-β-hydroxylase: A copper protein. *Journal of Biological Chemistry*, 1965, **240**, 552–554.

Fuller, R. W., & Hunt, J. M. Inhibition of phenylethanolamine-N-methyl transferase by its product, epinephrine. *Life Sciences (Part I)*, 1967, **6**, 1107–1112.

Glowinski, J., & Baldessarini, R. Metabolism of norepinephrine in the central nervous system. *Pharmacological Reviews*, 1966, **18**, 1201–1238.

Gordon, R., Spector, S., Sjoerdsma, A., & Udenfriend, S. Increased synthesis of norepinephrine in the intact rat during exercise and cold exposure. *Journal of Pharmacology and Experimental Therapeutics*, 1966, **153**, 440–447.

Griffith, J. J., Cavanaugh, J. H., & Oates, J. A. Psychosis induced by the administration of *d*-amphetamine to human volunteers. In D. H. Efron (Ed.), *Psychotomimetic drugs*. New York: Raven Press, 1970, pp. 287–298.

Hamburg, B. A., & Hamburg, D. A. Stressful transitions of adolescence: endocrine and psychosocial aspects. In L. Levi (Ed.), *Society, stress and disease, II: childhood and adolescence*. London: Oxford University Press, in press.

Hamburg, D. Genetics of adrenocortical hormone metabolism in relation to psychological stress. In J. Hirsch (Ed.), *Behavior-genetic analysis*. New York: McGraw–Hill, 1967, pp. 154–175.

Hamburg, D. A. A combined biological and psychosocial approach to the study of behavioral development. In A. Ambrose (Ed.), *Stimulation in early infancy*. New York: Academic Press, 1969, pp. 269–277.

Hamburg, D. A. (Ed.), *Psychiatry as a behavioral science*. Englewood Cliffs, New Jersey: Prentice Hall, 1970.

Hamburg, D. A., & Adams, J. E. A perspective on coping behavior. *Archives of General Psychiatry*, 1967, **17**, 277–284.

Hamburg, D. A., Adams, J. E., & Brodie, H. K. H. Coping behavior in stressful circumstances: some implications for social psychiatry. In A. H. Leighton (Ed.), *Further explorations in social psychiatry*. Basic Books, in press.

Hamburg, D. A., Hamburg, B. A., & Barchas, J. D. Anger and depression. In L. Levi (Ed.), *Parameters of emotion*. New York: Raven Press, in press.

Hamburg, D. A., & Kessler, S. A behavioral-endocrine-genetic approach to stress problems. In S. Pickett (Ed.), *Endocrine genetics. Memoirs of the Society for Endocrinology No. 15*. London: Cambridge University Press, 1967, pp. 249–270.

Hamburg, D., & Lunde, D. Relation of behavioral, genetic, and neuroendocrine factors to thyroid function. In J. Spuhler (Ed.), *Genetic diversity and human behavior*. Chicago: Aldine Press, 1967, pp. 135–170.

Hamburg, D. A., & Van Lawick-Goodall, J. Factors facilitating development of aggressive behavior in chimpanzees and humans. In W. W. Hartup & J. de Wit (Eds.), *Determinants and origins of aggressive behavior*. The Hague: Mouton, in press.

Hinde, R. A., & Spencer-Booth, Y. The study of mother-infant interaction in captive group-living rhesus monkeys. *Proceedings of the Royal Society of London. Series B*, 1968, **169**, 177–201.

Holmes, T. H., & Rahen, R. H. The social readjustment rating scale. *Journal of Psychosomatic Research*, 1967, **11**, 213–218.

Jensen, C. D., & Tolman, C. W. Mother-infant relationship in the monkey, *Macaca nemestrina*: The effect of brief separation and mother-infant specificity. *Journal of Comparative and Physiological Psychology*, 1962, **55**, 131.

Karczmar, A. G., & Scudder, C. L. Behavioral responses to drugs and brain catecholamine levels in mice of different strains and genera. *Federation Proceedings*, 1967, **26**, 1186–1191.

Kaufman, C., & Rosenblum, L. A. The reaction to separation in infant monkeys: anaclitic depression and conservation-withdrawal. *Psychosomatic Medicine*, 1967, **19**, 648–675.

Kessler, S., Ciaranello, R. D., Shire, J. G. M., & Barchas, J. D. Genetic variation in catecholamine synthesizing enyme activities. *Genetics*, 1971, **68** (1, Pt. 2), s 33. (Abstract)

Kessler, S., Ciaranello, R. D., Shire, J. G. M., & Barchas, J. D. Genetic variation in catecholamine-synthesizing enzyme activity. *Proceedings of the National Academy of Sciences (Washington, D.C.)*, 1972, **69**, 2448–2450.

Kety, S., & Schildkraut, J. J. Biogenic amines and emotion. *Science*, 1967, **156**, 21–30.

Kirshner, N., & Goodall, M. Formation of adrenaline from noradrenaline. *Biochimica et Biophysica Acta*, 1957, **24**, 658–659.

Klerman, G. L. Depression and adaptation. Paper presented at the NIMH Conference on Psychology and Depression, Warrington, Virginia, October 1971.

Kvetnansky, R., Weise, V. K., & Kopin, I. J. Elevation of adrenal tyrosine hydroxylase and phenylethanolamine-N-methyl transferase by repeated immobilization of rats. *Endocrinology*, 1970, **87**, 744–749.

Leff, M. J., Roatch, J. F., & Bunney, W. E. Environmental factors preceding the onset of severe depressions. *Psychiatry*, 1970, **33**, 293–311.

Lindemann, E. Symptomatology and management of acute grief. *American Journal of Psychiatry*, 1944, **101**, 141–148.

Lipton, M. A. Summary. In J. Barchas & E. Usdin (Eds.), *Serotonin and behavior*. New York: Academic Press, 1973, pp. 565–568.

Maas, J. W. Neurochemical differences between two strains of mice. *Science*, 1962, **137**, 621–622.

Maas, J. W. Neurochemical differences between two strains of mice. *Nature*, 1963, **197**, 255–257.

Maas, J. W., Dekirmenjian, H., & Jones, F. The identification of depressed patients who have a disorder of NE metabolism and/or disposition. In S. H. Snyder & E. Usdin (Eds.), *Frontiers in catecholamine research*. New York: Pergamon, 1973, pp. 1091–1096.

MacLean, P. D. The triune brain, emotion and scientific bias. In F. O. Schmitt, (Ed.), *The neurosciences: second study program*. New York: The Rockefeller University Press, 1970, pp. 336–349.

McConville, B. J., Boag, L. C., & Purohit, A. P. Mourning depressive responses of children in residence following sudden death of parent figure. *Journal of the American Academy of Child Psychiatry*, 1972, **11**, 341–364.

Mendlewicz, J., Fieve, R. R., Rainer, J. D., & Fleiss, J. L. Manic depressive illness: a comparative study of patients with and without a family history. *British Journal of Psychiatry*, 1972, **120**, 523–530. (a)

Mendlewicz, J., Fieve, R. R., Stallone, F., & Fleiss, J. L. Genetic history as a predictor of lithium response in manic-depressive illness. *Lancet*, 1972, **1**, 599–600. (b)

Mendlewicz, J., Fleiss, J. L., & Fieve, R. R. Evidence for X-linkage in the transmission of manic-depressive illness. *Journal of the American Medical Association*, 1972, **222**, 1624–1627. (c)

Milkovic, K., Deguchi, T., Winget, C., Barchas, J., Levine, S., & Ciaranello, R. The effect of maternal manipulation on the phenylethanolamine N-methyltransferase activity and corticosterone content of the fetal adrenal gland. *American Journal of Physiology*, 1974, **266**, 864–866.

Miller, F. P., Cox, R. H., Jr., & Maickel, R. P. Intrastrain differences in serotonin and norepinephrine in discrete areas of rat brain. *Science*, 1968, **162**, 463–464.

Mitchell, G. D., Harlow, H. F., Griffin, G. A., & Moller, G. W. Repeated separation in the monkey. *Psychonomic Science*, 1967, **8**, 5.

Molinoff, P. B., & Axelrod, J. Biochemistry of catecholamines. *Annual Review of Biochemistry*, 1971, **40**, 465–500.

Molinoff, P. B., Brimijoin, S., Weinshilboum, R., & Axelrod, J. Neurally mediated increase in dopamine-β-hydroxylase activity. *Proceedings of the National Academy of Sciences (Washington, D.C.)*, 1970, **66**, 453–458.

Molinoff, P. B., Weinshilboum, R., & Axelrod, J. A sensitive enzymatic assay for depamine-β-hydroxylase. *Journal of Pharmacology and Experimental Therapeutics*, 1971, **178**, 425.

Mueller, R. A., Thoenen, H., & Axelrod, J. Adrenal tyrosine hydroxylase: compensatory increase in activity after chemical sympathectomy. *Science*, 1969, **158**, 468–469.

Nagatsu, T., Levitt, M., & Udenfriend, S. Tyrosine hydroxylase: the initial step in norepinephrine biosynthesis. *Journal of Biological Chemistry*, 1964, **239**, 2910–2917.

Page, J. G., Kessler, R. M., & Vesell, E. S. Strain differences in uptake, pool size and turnover rate of norepinephrine in hearts of mice. *Biochemical Pharmacology*, 1970, **19**, 1381–1386.

Parkes, C. M. *Bereavement: studies of grief in adult life*. New York: International Universities Press, 1972.

Patrick, R. L., & Barchas, J. D. Regulation of catecholamine synthesis in rat brain synaptosomes. *Journal of Neurochemistry*, 1974, **23**, 7–15.

Paykel, E. S., Myers, J. K., Dienelt, M. N., Klerman, G. L., Lindenthal, J. J., & Pepper, M. P. Life events and depression. *Archives of General Psychiatry*, 1969, **21**, 753–760.

Pohorecky, L. A., Zigmond, M., Karten, H., & Wurtman, R. J. Enzymatic conversion of norepinephrine to epinephrine by the brain. *Journal of Pharmacology and Experimental Therapeutics*, 1969, **165**, 190–195.

Reis, D. J. Central neurotransmitters in aggression. In S. Frazier (Ed.), *Association for Research in Nervous and Mental Diseases Symposium on Aggression*. Baltimore: Williams & Wilkins, in press.

Reis, D. J., & Fuxe, K. Brain norepinephrine: evidence that neuronal release is essential for sham rage behavior following brainstem transection in cat. *Proceed-*

ings of the National Academy of Sciences (Washington, D.C.), 1969, **64**, 108–112.

Reis, D. J., & Gunne, L.-M. Brain catecholamines: relation to the defense reaction evoked by amygdaloid stimulation in cat. *Science*, 1965, **149**, 450–451.

Rimoin, D., & Schimke, R. *Genetic disorders of the endocrine glands.* St. Louis: C. V. Mosby, 1971.

Rosenblum, L. A., & Kaufman, I. C. Variations in infant development and response to maternal loss in monkeys. *American Journal of Orthopsychiatry*, 1968, **38**, 418–426.

Roth, R. H., Stjärne, L., & von Euler, U. S. Factors influencing the rate of norepinephrine biosynthesis in nerve tissues. *Journal of Pharmacology and Experimental Therapeutics*, 1967, **158**, 373–377.

Schildkraut, J. J., Keeler, B. A., Papousek, M., & Hartmann, E. MHPG excretion in depressive disorders: relation to clinical subtypes and desynchronized sleep. *Science*, 1973, **181**, 762–764.

Schildkraut, J. J., & Kety, S. S. Biogenic amines and emotion. *Science*, 1967, **156**, 21–30.

Schlesinger, K., Boggan, W. O., & Freedman, D. X. Genetics of audiogenic seizures: I. relation to brain serotonin and norepinephrine in mice. *Life Sciences (Part I)*, 1965, **4**, 2345–2351.

Seay, B., Hansen, E., & Harlow, H. F. Mother-infant separation in monkeys. *Journal of Child Psychology and Psychiatry*, 1962, **3**, 123–132.

Seay, B., & Harlow, H. F. Maternal separation in the rhesus monkey. *Journal of Nervous and Mental Disease*, 1965, **140**, 434–441.

Sedvall, G. C., & Kopin, I. J. Influence of sympathetic denervation and nerve impulse activity on tyrosine hydroxylase in the rat submaxillary gland. *Biochemical Pharmacology*, 1967, **16**, 39–46.

Sedvall, G. C., Weise, V. K., & Kopin, I. J. The rate of norepinephrine synthesis measured *in vivo* during short intervals; influence of adrenergic nerve impulse activity. *Journal of Pharmacology and Experimental Therapeutics*, 1968, **159**, 274–282.

Shire, J. G. M. Genetics and the study of adrenal and renal function in mice. In C. Gual & F. J. G. Ebling (Eds.), Progress in endochrinology. *Excerpta Medica. International Congress Series*, 1969, No. 184, pp. 328–332.

Shire, J. G. M. Genetic variation in adrenal structure: quantitative measurements on the cortex and medulla in hybrid mice. *Journal of Endocrinology*, 1970, **48**, 419–431.

Snyder, S. H. Catecholamines in the brain as mediators of amphetamine psychosis. *Archives of General Psychiatry*, 1972, **27**, 169–179.

Snyder, S. H. Amphetamine psychosis: a "model" schizophrenia mediated by catecholamines. *American Journal of Psychiatry*, 1973, **130**, 61–67.

Spector, S., Gordon, R., Sjoerdsma, A., & Udenfriend, S. End-product inhibition of tyrosine hydroxylase as a possible mechanism for regulation of norepinephrine synthesis. *Molecular Pharmacology*, 1967, **3**, 549–555.

Spencer-Booth, Y., & Hinde, R. A. The effects of separating rhesus monkey infants from their mothers for six days. *Journal of Child Psychology and Psychiatry*, 1967, **7**, 179–197.

Stempfel, R. S., & Tomkins, G. M. Congenital virilizing adrenocortical hyperplasia (the adrenogenital syndrome). In J. B. Stanbury, J. B. Wyngaarden, D. S.

Fredrickson (Eds.), *The metabolic basic of inherited disease*. New York: McGraw-Hill, 1966, pp. 635–664.

Stolk, J. M., Conner, R. L., Levine, S., & Barchas, J. D. Brain norepinephrine metabolism and shock-induced fighting behavior in rats: differential effects of shock and fighting on the neurochemical response to a common footshock stimulus. *Journal of Pharmacology and Experimental Therapeutics*, 1974, **190**, 193–209.

Sudak, H. S., & Maas, J. W. Central nervous system serotonin and norepinephrine localization in emotional and non-emotional strains in mice. *Nature*, 1964, **203**, 1254–1256. (a)

Sudak, H. S., & Maas, J. W. Behavioral-neurochemical correlation in reactive and nonreactive strains of rats. *Science*, 1964, **146**, 418–420. (b)

Thoenen, H., Mueller, R. A., & Axelrod, J. Trans-synaptic induction of adrenal tyrosine hydroxylase. *Journal of Pharmacology and Experimental Therapeutics*, 1969, **169**, 249–254.

Thoenen, H., Mueller, R. A., & Axelrod, J. Neuronally dependent induction of adrenal phenylethanolamine-N-methyltransferase by 6-hydroxydopamine. *Biochemical Pharmacology*, 1970, **19**, 669–674.

Tinklenberg, J. R. A clinical view of amphetamines. *American Family Physician*, 1971, **4**, 82–86.

Van Lawick-Goodall, J. The behavior of free-living chimpanzees in the Gombe Stream area. *Animal Behavior. Monographs*, 1968, **1**(3), 161–311.

Van Lawick-Goodall, J. The behavior of chimpanzees in their natural habitat. *American Journal of Psychiatry*, 1973, **130**, 1–12.

Vernikos-Danellis, J., Ciaranello, R., & Barchas, J. Adrenal epinephrine and phenylethanolamine N-methyl transferase (PNMT) activity in the rat bearing a transplantable pituitary tumor. *Endocrinology*, 1968, **83**, 1357–1358.

Vogt, M. The concentration of sympathin in different parts of the central nervous system under normal conditions and after the administration of drugs. *Journal of Physiology (London)*, 1954, **123**, 451–481.

Weiner, N., & Rabadjija, M. The effect of nerve stimulation on the synthesis and metabolism of norepinephrine in the isolated guinea-pig hypogastric nerve-vas deferens preparation. *Journal of Pharmacology and Experimental Therapeutics*, 1968, **160**, 61–71.

Weinshilboum, R., & Axelrod, J. Dopamine-β-hydroxylase activity in the rat after hypophysectomy. *Endocrinology*, 1970, **87**, 894–899.

Weinshilboum, R. M., Raymond, F. A., Elueback, L. R., & Weidman, W. H. Dopamine-β-hydroxylase activity in serum. In S. H. Snyder & E. Usdin (Eds.), *Frontiers in catecholamine research*. New York: Pergamon, 1973, pp. 1115–1121.

Welch, B. L., & Welch, A. S. Aggression and the biogenic amine neurohumors. In S. Garattini & E. B. Sigg (Eds.), *Aggressive behavior*. New York: John Wiley & Sons, 1969, pp. 188–202. (a)

Welch, B. L., & Welch, A. S. Fighting: preferential lowering of norepinephrine and dopamine in the brainstem, concomitant with a depletion of epinephrine from the adrenal medulla. *Communications in Behavioral Biology*, 1969, **3**, 125–130. (b)

Welch, B. L., & Welch, A. S. Sustained effects of brief daily stress (fighting) upon brain and adrenal catecholamines and adrenal, spleen and heart weights of mice.

Proceedings of the National Academy of Sciences (Washington, D.C.), 1969, **64**, 100–107. (c)

Winokur, G., Morrison, J., Clancy, J., & Crowe, R. The Iowa 500: familial and clinical findings favor two kinds of depressive illness. *Comprehensive Psychiatry*, 1973, **14**, 99–106.

Wurtman, R. J., & Axelrod, J. Control of enzymatic synthesis of adrenaline in the adrenal medulla by adrenal cortical steroids. *Journal of Biological Chemistry*, 1966, **241**, 2301–2305.

Wyatt, R., Portnoy, B., Kupfer, D., Snyder, F., & Engelman, K. Resting plasma catecholamine concentrations in patients with depression and anxiety. *Archives of General Psychiatry*, 1971, **24**, 65–70.

II GENETIC STUDIES OF CRIMINALITY AND PSYCHOPATHY

4 EXTRA CHROMOSOMES AND CRIMINALITY

Seymour Kessler

Criminal behavior has the characteristics expected of genetically determined traits. A familial concentration tends to occur. Also, concordance rates for criminality among monozygotic twins are generally higher than among dizygotic twin pairs (Rosenthal 1970). The evidence, however, is equivocal largely because the effects of psychosocial and cultural factors have not been separated from those of genetic factors (Mulvihill & Tumin 1969).

Over the past decade, interest in the behavioral consequences of chromosomal disorders has led to the renewed study of possible hereditary contributions to criminality. In 1965 Patricia Jacobs and her colleagues published the results of a chromosomal survey carried out in a maximum security State Hospital at Carstairs, Scotland (Jacobs, Brunton, Melville, Brittain, & McClemont 1965). Among 315 inmates studied, 9 XYY men were detected. In subsequent publications (Price, Strong, Whatmore, & McClemont 1966; Price & Whatmore 1967a, 1967b; Jacobs, Price, Court-Brown, Brittain, & Whatmore 1968), these XYY individuals were characterized as suffering from a severe personality disorder marked by impulsivity and an uncontrollable drive toward antisocial behavior. On the average, they were taller than their chromosomally normal fellow inmates, and their criminal activities had started earlier. Also, there was virtually no predisposing family history of criminality to account for their life of crime and eventual incarceration. Inferentially then, it was the extra Y chromosome that was responsible for bringing them into conflict with the law. Lastly, it was pointed out that the 9 XYY men accounted for nearly 3 percent of the Carstairs Hospital population, whereas the rate of XYY males among the newborn was only about 0.1 percent. Thus XYY individuals appeared to be at considerably greater than average risk for institutionalization.

I would like to address myself to several issues raised by the research on the relationship between an extra Y chromosome and criminality. Specifically, I plan to consider some of the factors that might account for the

Seymour Kessler, Ph.D., Department of Psychiatry, Stanford University School of Medicine, Stanford, California 94305.

apparently high rate of XYY males in institutions, the possible magnitude of their increased risk for institutionalization, and the implications of the latter, particularly for genetic counseling.

Since the appearance of the initial report from the Jacobs' group, over forty other chromosomal surveys of institutionalized male populations have been published. The most extensive compilation of these data has been provided by Hook (1973), who divided the surveys according to the nature of the institutional setting into (1) a mental group, consisting of retarded, disturbed, psychotic, alcoholic, or epileptic individuals; (2) a penal group, in which punitive or security arrangements were present; (3) a mental-penal group, in which features of both were present (e.g., a hospital for the criminally insane). Hook and others have suggested that a definite association exists between the XYY karyotype and the presence in a mental-penal setting. However, it needs to be kept in mind that the various settings do not represent homogeneous groupings of institutions. Within each setting, individuals in different institutions may differ on such variables as age, ethnic background, and length of institutionalization. Within different institutions, differing proportions of individuals with diagnoses of mental subnormality and mental illness may be present. Also, a diversity of admissions and discharge criteria exist. These, in turn, are influenced by local and regional legal requirements and policies that may be only incidentally related to the behavior of some of the individuals in that setting. For example, individuals may be committed to a mental-penal institution by court order, but they may also be transferred there for administrative or other reasons. It would be a mistake to use the label providing a description of the general nature of an institution as a description of the pre- or post-admission behavior of the inmates of that institution. Also, it is not at all clear that the preadmission offenses of males with chromosomal disorders in mental-penal institutions differ in any major way from those of similarly age-matched individuals in penal institutions. For these several reasons, I will consider the overall results of the institutional surveys rather than data separated by divisions based on setting.

Examination of the chromosomal surveys in the penal, special security and mental hospitals of North America and Europe reveals that of 13,522 males studied at least 147 XYY individuals have been detected, yielding a rate of 10.9 per 1,000. The rate of this chromosomal disorder generally found in the major surveys among newborn males is approximately one per 1,000 (Table 1). Thus, the institutional rate appears to represent a remarkable increase of XYY males over the rate expected on the basis of the mean newborn incidence.

Although it is tempting to ascribe the institutional rate of the XYY chromosomal disorder to the presence of the extra Y chromosome, several qualifications need to be considered. First, factors aside from the extra Y chromosome might account for the apparently high rate of XYY males in

Table 1. XYY and XXY Disorders Detected in Chromosomal Surveys among Newborn Males (data from Hook 1973 and Lubs 1972)

	N	XYY	XXY
Boston	4,703	3	4
Edinburgh	4,831	7	6
London, Ontario	1,066	4	1
New Haven	2,176	3	4
Winnipeg	3,468	1	2
Total	16,244	18	17

institutional settings. The increased risk for institutionalization is not specific for males with an extra Y chromosome; XXY males also appear to be at higher than average risk for institutionalization. The XXY disorder occurs among newborn males, also at a rate of about one per 1,000 (Table 1). In the chromosomal surveys cited above, 104 XXY males were found in various institutions, yielding a rate of 7.7 per 1,000. If the ratio of XYY and XXY males detected in the newborn surveys is compared to the overall ratio of males with these two disorders in institutions, a surprising similarity of the ratios emerges. Among the newborn surveys, the XYY:XXY ratios range between 0.5 and 4.0 with a pooled mean of 1.1, whereas, in the institutions the range of the ratios is between 0.3 and 9.0, with a pooled mean of 1.4. The pooled means are not significantly different. Thus both males with an extra X or an extra Y chromosome are at apparently higher than average risk for institutionalization. Furthermore, the risk is relatively constant for both disorders suggesting that a chromosomal disorder per se or factors independent of the supernumerary chromosomes are responsible for the institutionalization of XYY and XXY individuals. The possibility is raised that these factors might be the same for both chromosomal disorders.

Second, the difference between the institutional and newborn rates of the XYY disorder may be more apparent than real. Newborn rates of chromosomal disorders are subject to both short- and long-range variations as a consequence of multiple factors, including season of the year at time of conception, rubella epidemics, patterns of drug use, exposure to radiation, changing marriage patterns, fertility rates, and so on (Harris & Robinson 1971). Also, there are variations in the technical procedures used by different laboratories in the determination of individual karyotypes. These differences include, among other things, the number of cells counted and the rigor with which mosaicism is ruled out. The fact that most chromosomal disorders occur at low base rates suggests that small errors in karyotyping procedures may have important effects on published incidence figures (Shah 1970).

Third, the comparison of the prevalence of a chromosomal disorder among institutionalized males to the incidence among newborn males assumes that the latter population is matched to the former on social class, ethnic composition, and other variables. Insufficient evidence is available to determine whether this assumption is correct. If the rate of chromosomal nondisjunction is greater among some socioeconomic groups than among others, as some available data seem to suggest (Robinson & Puck 1967), then a more adequate matching of institutional and newborn samples would be needed before concluding that differential rates exist (Kessler & Moos 1970).

Several investigators have suggested that the contribution of socioeconomic variables to the institutionalization of XYY males is of little or no importance. Hook (1973) has implied that XYY males are drawn from relatively higher social classes than XY inmates. Although the problem needs greater attention in future research, currently available data suggest that a considerable proportion of XYY males in institutional settings derive from lower social classes. For example, Casey (1969) studied the family backgrounds of 33 XYY and 150 chromosomally normal fellow inmates in two special security hospitals (Rampton and Moss Side) in England and found that 18 of 26 (69.2 percent) fathers of the XYY inmates and 66 of 113 (58.4 percent) control fathers came from social classes IV and V. Also, where the information on family background is provided, it is abundantly clear that institutionalized XYY males are likely to have been born into and reared in environmental circumstances conducive to institutionalization. Casey found that

1) although the average family size in the general population was about 2.2, among the XYY and control inmates it was 5.2 and 4.4 respectively;

2) although the illegitimacy rate in the general population was about 5 percent, among the XYY and control inmates it was 15.2 and 14.6 percents respectively;

3) the proportion of XYY cases showing evidence of parental absence for 6 or more months before age 6 was 51.5 percent, whereas, among the control subjects it was 32.7 percent;

4) the proportion of XYY subjects with a history of parental and/or sibling conviction was 27.3 percent for the XYY cases and 18.7 percent for the controls, a finding inconsistent with those of Price and Whatmore (1967a);

5) the combined rate of parental absence before age 6, and/or familial conviction or mental disorder was about 72.7 percent for the XYY subjects and 53.3 percent for the controls.

In an update of the earlier survey, Casey et al. (1973) report a less striking downward shift in parental social class among the inmates at Rampton and Moss Side Hospitals. Nonetheless, they state that ". . . fewer fathers than expected belong to social classes 1 and 2."

From the literature and from his own clinical research, Money (1970) compiled a group of 25 XYY males for whom sufficient background information was provided. Of these individuals, 60 percent showed a family history rated as conducive to psychopathology. From the published institutional surveys, I have compiled an additional 54 cases of XYY males for whom information on family background was given in sufficient detail. These are shown in Table 2.

Individuals with a family history of various combinations of criminality, psychiatric disorder, parental separation or death, illegitimacy, alcoholism, child abuse, or pregnancy and birth complications are listed as having a positive family history conducive to institutionalization. Table 2 shows that over 75 percent of the XYY males derive from families with a history of criminality and/or other psychosocial disturbance. These data strongly suggest that the circumstances into which XYY males are born may play a major role in their later institutionalization. Moreover, they suggest that it is premature, if not unwarranted, to ascribe a causal role for behavioral deviancy to the XYY chromosomal disorder since the effects of associated environmental variables have not been separated from the effects of the disorder itself.

Casey's data (1969) also suggest that a downward shift of parental social class and a high rate of familial conviction and/or psychosocial disturbance may also be present among institutionalized XXY subjects. Thus factors associated with socioeconomic status may be a common denominator promoting an apparently constant risk for institutionalization among XYY and XXY males. Analysis of Casey's data suggests that knowing the parental social class of an individual may be two or more times more effective in predicting future behavioral deviancy than knowing his karyotype.*

The possibility remains, however, that even if socioeconomic and other factors are taken into account, an increased risk for institutionalization may still be present among XYY males. Although an accurate estimate of

*From the distribution of socioeconomic and other related variables in the general population and among institutionalized individuals, the conditional probability may be calculated that given the presence or absence of a characteristic, C, deviance would be expected. The increased risk of deviancy, R, associated with C can be computed from the formula:

$$R = \frac{\text{Pr}\left\{C/\text{deviant}\right\}}{\left[1-\text{Pr}\left\{C/\text{deviant}\right\}\right]} \frac{\left[1-\text{Pr}\left\{C\right\}\right]}{\text{Pr}\left\{C\right\}} \quad \frac{\text{Pr}\left\{\text{deviant}/C\right\}}{\text{Pr}\left\{\text{deviant}/\text{not }C\right\}}$$

Thus, if C is not associated with behavioral deviance, $R = 1$; $R > 1$ if individuals with C are more likely to be deviant than individuals without C. Let us say that C represents having the XYY karyotype, then from Casey's data, $\text{Pr}\left\{C/\text{deviant}\right\} = 0.0028$ and $\text{Pr}\ C = 0.0014$; $R = 1.97$. This could be interpreted as meaning that the risk for deviancy among XYY males is twice as high as among those who are chromosomally normal. However, if C represents illegitimacy, then $R = 3.09$ and if C represents having a father of social class V (unskilled laborers), $R = 4.55$. I thank Dr. H. Kraemer for calling this to my attention.

Table 2. Family History Status (see Text) of XYY Males Detected in
Chromosomal Surveys

		Positive	Negative	Total
Court Brown et al.	(1968)	1	0	1
Akesson et al.	(1969)	2	1	3
Casey	(1969)	24	9	33
Marinello et al.	(1969)	4	0	4
Nielsen et al.	(1969)	2	0	2
Baker et al.	(1970)	6	0	6
Matsaniotis et al.	(1970)	1	1	2
Tsuboi	(1970)	1	2	3
	Total	41	13	54
		(75.9%)	(24.1%)	

the magnitude of this risk is not currently possible, a rough estimate can be calculated for at least one population, namely, that of Scotland. In 1971 Patricia Jacobs and her colleagues published the results of a chromosomal survey of the penal institutions and approved schools in Scotland (Jacobs, Price, Richmond, & Ratcliff 1971). This study is distinguished by at least three features: (1) careful attention to the problem of sampling; a large group of inmates was studied and the sampling was probably representative of the entire penal and approved school population of a restricted geographic region over a specified period of time; (2) a single laboratory carried out all of the karyotype determinations, thus laboratory procedures and techniques were relatively constant over the course of the study; (3) the same laboratory determined the rates of chromosomal disorders among the newborn male population of Scotland, at least those of the greater Edinburgh area. How representative this latter group is of the institutionalized population is not known, but, presumably, it is more representative of Scottish penal and approved school inmates than other newborn groups.

The rate at which the XYY disorder occurs among Scottish newborn males is 1.4 per 1,000 (Table 1). Since there are about 1,816,100 males in Scotland 15 or more years of age, 2,543 of these would be expected to have the XYY karyotype. This calculation assumes that no differential mortality of XYY individuals occurs relative to chromosomally normal males. From the survey of males in the penal and approved schools, it can be shown that the expectancy of institutionalization among XY males would be 0.14 percent, whereas, among XYY males the expectancy would be 0.28 percent, twice as high as among chromosomally normal males in Scotland. However, an elevated risk of this magnitude is small if one considers the chances of noninstitutionalization in this group. These are 99.86 and 99.73 percent for chromosomally normal and XYY males respectively.

The Carstairs Hospital represents a mental-penal institution. Jacobs et al. have pointed out that the 9 XYY inmates at this hospital represent 0.36 percent of the expected XYY population of Scotland, whereas, the total Carstairs sample represents only 0.02 percent of the general adult male population. Thus XYY males appear to show a twenty times greater risk for institutionalization in a mental-penal institution than chromosomally normal males. However, even with an increase of risk of this magnitude, 99.6 percent of Scottish XYY males would not be expected to be institutionalized at Carstairs Hospital. In fact, over 99 percent of XYY males would not be expected to be institutionalized in any institution in Scotland. These calculations suggest that by informing parents of an XYY child that he is twenty times more likely to be institutionalized than a chromosomally normal male, we may be misleading them to believe that the risk for behavioral deviancy is of a far greater magnitude than the facts warrant.

The fact that the overwhelming majority of XYY males do not appear to be in institutional settings raises intriguing questions. Where are these individuals and what are the factors that permit their apparently successful adaptation to societal mores and regulations? Answers to these questions might provide us with a greater understanding of the factors that promote the behavioral deviancy of some of these individuals and with clues as to strategies that might prevent or mitigate such behavior. One of the possible factors that may have adaptive consequences for at least some XYY males is their apparent tendency to greater than average stature. Evidence exists suggesting that taller adolescents tend to show greater leadership, popularity, social success, and good social adjustment more frequently than relatively shorter adolescents. Thus, some XYY males might be predisposed toward social assertiveness, the attainment of economic success, and toward making outstanding social adjustments and contributions as a consequence of their increased stature. It cannot be simply assumed that chromosomal disorders must promote maladaptive consequences. In fact, the persistent focus on abnormality may seriously compromise the objectivity of investigators (Kessler & Moos 1973).

With respect to the issue of genetic counseling, if we inform parents that their child has a substantial risk for eventual institutionalization, are we not thereby promoting a self-fulfilling prophecy? If genetic counseling is preventive medicine, then considerable attention needs to be given to the quality of information given to the parents of children with chromosomal disorders. If is insufficient to inform them that a twenty-fold increase in risk for deviancy exists, since, in absolute terms, the chances for institutionalization among XYY males appears to be small. On the whole, these individuals do not appear to be destined to a life of crime. Information reinforcing this view needs to be strongly conveyed during the course of counseling.

References

Akesson, H. O., Forssman, H., & Wallin, L. Gross chromosomal errors in tall men admitted to mental hospitals, *Acta Psychiatrica Scandinavica,* 1969, **45**, 37–46.

Baker, D., Telfer, M. A., Richardson, C. E., & Clark, G. R. Chromosome errors in men with antisocial behavior. *Journal of the American Medical Association,* 1970, **214**, 869–878.

Casey, M. D. The family and behavioural history of patients with chromosome abnormality in the special hospitals of Rampton and Moss Side. In D. J. West (Ed.), *Criminological implications of chromosome abnormalities.* Cambridge: Institute of Criminology, 1969, pp. 49–60.

Casey, M. D., Blank, C. E., McLean, T. M., Kohn, P., Street, D.R.K., McDougall, J. M., Gooder, J., & Platts, J. Male patients with chromosome abnormality in two state hospitals. *Journal of Mental Deficiency Research,* 1973, **16**, 215–256.

Court Brown, W. M., Price, W. H., & Jacobs, P. A. Further information on the identity of 47, XYY males. *British Medical Journal,* 1968, **2**, 325–328.

Harris, J. S., & Robinson, A. X chromosome abnormalities and the obstetrician. *American Journal of Obstetrics and Gynecology,* 1971, **109**, 574–583.

Hook, E. B. Behavior implications of the human XYY genotype. *Science,* 1973, **179**, 139–150.

Jacobs, P. A., Brunton, M., Melville, M. M., Brittain, R. P., & McClemont, W. F. Aggressive behaviour, mental sub-normality and the XYY male. *Nature,* 1965, **208**, 1351–1352.

Jacobs, P. A., Price, W. H., Court Brown, W. N., Brittain, R. P., & Whatmore, P. A. Chromosome studies on men in a maximum security hospital. *Annals of Human Genetics,* 1968, **31**, 339–347.

Jacobs, P. A., Price, W. H., Richmond, S., & Ratcliff, R. A. W. Chromosome surveys in penal institutions and approved schools. *Journal of Medical Genetics,* 1971, **8**, 49–58.

Kessler, S., & Moos, R. H. The XYY karyotype and criminality: a review. *Journal of Psychiatric Research,* 1970, **7**, 153–170.

Kessler, S., & Moos, R. H. Behavioral aspects of chromosomal disorders. *Annual Review of Medicine,* 1973, **24**, 89–102.

Lubs, H. A. Neonatal cytogenetic surveys. In S. W. Wright, B. F. Crandall, & L. Boyer (Eds.), *Perspectives in cytogenetics.* Springfield: C. C Thomas, 1972, pp. 297–304.

Marinello, M. J., Berkson, R. A., Edwards, J. A., & Banneman, R. M. A study of the XYY syndrome in tall men and juvenile delinquents. *Journal of the American Medical Association,* 1969, **208**, 321–325.

Matsaniotis, N., Tsenghi, C., Metaxotou-Stavridaki, C., Economou-Mavrou, C., & Bilalis, P. The XYY syndrome in young Greek detainees. *Helvetica Paediatrica Acta,* 1970, **25**, 253–257.

Money, J. Impulse, aggression and sexuality in the XYY syndrome. *St. John's Law Review,* 1970, **44**, 220–235.

Mulvihill, D. J., & Tumin, M. M. Crimes of violence; a staff report submitted to the National Commission on the Causes and Prevention of Violence. Washington, D.C.: Superintendent of Documents, U.S. Government Printing Office, 1969.

Nielsen, J., Sturup, G., Tsuboi, T., & Romano, D. Prevalence of the XYY syndrome in an institution for psychologically abnormal criminals. *Acta Psychiatrica Scandinavica*, 1969, **45**, 383–401.

Price, W. H., Strong, J. A., Whatmore, P. B., & McClemont, W. F. Criminal patients with XYY sex-chromosome complement. *Lancet*, 1966, **1**, 565–566.

Price, W. H., & Whatmore, P. B. Behaviour disorders and pattern of crime among XYY males identified at a maximum security hospital. *British Medical Journal*, 1967, **1**, 533–536.(a)

Price, W. H., & Whatmore, P. B. Criminal behaviour and the XYY male. *Nature*, 1967, **213**, 815.(b)

Robinson, A., & Puck, T. T. Studies on chromosomal nondisjunction in man. II. *American Journal of Human Genetics*, 1967, **19**, 112–129.

Rosenthal, D. *Genetic theory and abnormal behavior.* New York: McGraw–Hill, 1970.

Shah, S. A. Report on the XYY chromosomal abnormality. Public Health Service Publication No. 2103. Chevy Chase, Md.: National Institute of Mental Health, 1970.

Tsuboi, T. Crimino-biologic study of patients with the XYY syndrome and Klinefelter's syndrome. *Humangenetik*, 1970, **10**, 68–84.

5 CYTOGENETIC AND DERMATOGLYPHIC STUDIES IN SEXUAL OFFENDERS, VIOLENT CRIMINALS, AND AGGRESSIVELY BEHAVED TEMPORAL LOBE EPILEPTICS

Lawrence Razavi

Sex chromosomal abnormalities are found more commonly among aggressively behaved and sexually disturbed prison inmates than in the general population (Tables 1 and 2). The excess varies between ten- and fifty-fold (2 percent to 10 percent versus 0.2 percent in newborns); and though prominence has been given to the XYY karyotype in reports of these surveys, a variety of chromosomal constitutions involving excess or deficiency of either X or Y chromosome are found (Table 3) in groups of either male or female (putative) sex. Furthermore, mosaic cases occur in which varying numbers of cells are affected: these proportions may change over time, and their tissue distributions in the body may differ from one individual to another (Razavi 1970, 1972). In hospital studies, on the other hand, it is often observed that sexual aneuploids attending for fertility, endocrine, and neurological problems are psychiatrically disturbed. They may have a variety of psychosexual conflicts and neuroses and are occasionally impulsive in behavior, though they need not be violent.

There is still some uncertainty about the interpretation of these data:

1. The rate of aneuploidy in young, normal adults is not known, though there is an increasing incidence of chromosomal mutation in aging populations, particularly above the age of sixty. Studies to date have relied on comparisons between young adult deviants and newborn infants, so that differences between the two are sometimes regarded as over-estimates or even spurious. On the other hand, the differences in rate between carefully ascertained institutionalized groups and available controls are hard to encompass within what is known of normal age-dependent variability,

Lawrence Razavi, M. B., B.S. Human Development Research Unit, Park Hospital for Children, Oxford, England.

Table 1. Antisocial Behavior and Aneuploidy

Population	X excess (%)	Y excess (%)	Autosome* excess (%)
Forssman & Hambert 1963			
760 aggressive mental defectives	2.0	0	0
1,625 mental defectives	0.6	0	0
Wegmann & Smith 1963			
505 juvenile delinquents	0	0	0
813 repeated male offenders	0.2	0	0
Jacobs et al. 1965			
197 male criminals		3.0	0.2
Casey et al. 1966a,b			
10,685 mental defectives	0.84	0.1	0.1
942 aggressive mental defectives	2.2		
Welch et al. 1967			
21 tall defective delinquents	5.0	–	–
Goodman et al. 1967			
100 tall criminals	2.0	2.0	0
Telfer et al. 1968			
129 tall criminals	5.0	4.0	0
Hunter 1968			
1021 juvenile delinquents	2.0	3.0	–
Melnyk et al. 1969			
79 tall sexual offenders	–	8.0	–
Razavi 1969			
83 sexual offenders	5.0	4.0	1.0

*Autosomes: non-sex chromosomes.

and the predictability and consistency of the yield among diversely selected adult populations of either sex, whose major common characteristic
is severe and dangerous behavior, makes it likely that the phenomenon
is real and worth further investigation.

2. Most studies examine single tissues, usually lymphocytes from peripheral blood, less commonly skin. However, these tissues are not a necessary
part of the neuroendocrine centers involved in behavior. Since many examples are known of aneuploidy affecting localized areas only (e.g., a
cleft palate or a tumor), cytogenetical anomaly in nonneural tissues may
not be regarded as a fully convincing demonstration of a physiological link
between genetic disorder and behavioral illness. At the same time,
neuroendocrine disturbance and brain dysplasia are well-documented in
chromosomal anomalies (Tables 4, 5), and the concurrence of epilepsy with
sexual aneuploidy is greater than that expected by chance, particularly if
the epilepsy is associated with mental illness. The association is stronger

Table 2. Cytogenetics of Institutionalized Male Sexual Offenders (N=65) (Kaplan 1965, 1974)

Major Mosaics (→ 100% cells aneuploid)

Case	Study #	Age	HT″	I.Q. WAIS	Crime classification*	Karyotypes	Clinical appearance	Plasma testosterone mg/100ml
1	42	26	60	67	Fixated pedophile	XXY	Small genitalia but fertile	0.57
2	67	38	71	69	Compensatory rapist	XXY	Testicular atrophy, diabetes	
3	69	36	65	97 77.7	Fixated pedophile	XXY 2X,1Y	Urethral cyst	0.42
Intermediate mosaics (> 20% aneuploid)								
4	91	22	65	118	Fixated pedophile	XY/XYY	Normal	1.35
5	63	24	66	75	Impulsive rapist	XY/XO	Normal	
6	75	24	72	122 105	Compensatory rapist	XY/XXY 1X,1Y,1O	Neurofibromatosis	
Minor mosaics (10–20% aneuploid)								
7	25	60	66	98	Compensatory rapist	XY/XXY	Normal	
8	26	51	71	79	Regressed pedophile	XY/XYYq+	Normal	0.49
9	62	26	67	110	Sex-aggression fusion rapist	XY/XO/XXY	EKG: PR interval 0.4 sec.	
10	37	53	64	107	Regressed pedophile	XY/XYY	Normal	
11	51	30	67	110 100.8	Displaced rapist	XY/XYY 2X,3Y,1O	Normal	
Insignificant mosaics (less than 10%; if more than 10% only single samples: within normal variation or limits of experimental error)								
	10	39	67	105	Displaced rapist	XY/XXY	Normal	0.80
	13	34	67	99	Compensatory rapist	XY/XYY	Normal	0.34
No cytogenetic anomalies detected								
	81	57	69	73	Fixated pedophile	XY	Normal	0.80
	2	29	68	89	Impulsive rapist	XY	Normal	0.69
	112	28	71	98	Displaced rapist	XY	Normal	0.70

*Social characteristics: early or middle birth order, 4–6 siblings.
Offenses: first in adolescence, none previous or over four previous.
Upbringing: small urban, conservative family attitudes, irregular schooling, sporadic jobs.
Rare: previous hospitalization, drunkenness, nonsexual crime, use of dangerous weapon.

Table 3. Cytogenetical Surveys of Behavioral Disorders (Razavi 1969, 1970, 1974)

	Male				Female	
	Sexual offenders		Federal inmates (biased)	Temporal lobe epileptics	State penitentiary	Temporal lobe epileptics
Series	I	II				
Number	82	17	68	18	15	4
Karyotypes	12	1	7	2	1	0
	2XXY	XXY	XXY	XXY	XXX	
	XYY		2XYY	XYY		
	2XY/XXY		XY/XXY			
	4XY/XYY		XY/XYY			
	XY/XO		XY/XO			
	XY/XO/XYY		XY/XO/XYY			
	t(A2q–A1q+)					
Percent	14.6%	5.9%	10.3%	11.1%	6.7%	
Significant dermatoglyphics	A,LU ↑ TFRC ↓		A,LU ↑ TFRC ↓	small LU ↑ TFRC ↓	small LU ↑ t'u	small LU ↑ t'u

Table 4. Sex Hormone Levels According to Karyotype

| Subjects | Urinary sex hormone excretion/24 hrs.—international units | | | | |
	Testosterone	FSH	LH	FSH/LH ratio	References
XYY	81±11 mg	15±1.5 IU	77±1.7 IU	0.2	Ismail et al.
Aggressive males	69±9 mg				1968
Normal males	52±2 mg				
XXY	36±4 mg				
	Serum and urinary sex hormone levels—mouse units and weight				
XXY	483 mg	50–100 MU	17 MU		Shapiro
Mental defective	–	–	9–22 MU		1970
	Serum hormone levels—mouse units and weight				
XXY	340–480 mg	250–700 mg/ml	10–52 MU		Santen et al.
Normal males	280–144 mg	150–610 mg/ml	44–23 MU		1970

in males than females, in mental institutions than in hospital clinics, and in groups characterized first by karyotype than in those initially ascertained on the basis of the neurological disorder (Tables 6, 7). It seems possible, therefore, that the association between aneuploidy and rage-prone psychoneurosis is in fact mediated by a special effect of the chromosomal anomaly on brain development, and that this may be confined to small foci in inaccessible parts of the limbic and neuroendocrine systems.

3. Large surveys usually study a single sample from each individual. It is known, however, that in mosaics the number of affected cells may vary not only from site to site but also over time, especially in high-turnover tissues such as blood (Razavi 1972): this is what occurs in myeloid leukemia, and it probably occurs also in tissues showing age-dependent maturation, at times in life when they undergo special proliferation. Even if appropriate tissues were available, the difficulty is that cases with minimal aneuploidy at the time of sampling may escape detection; and those with an initially high level, which subsequently declines, may be wrongfully regarded as a major genetic hazard. Repetitive testing is therefore necessary although it is expensive and time-consuming.

Ascertainment and Sampling

These difficulties are largely ones of ascertainment and sampling, the former depending on definitions of abnormal stereotypes of behavior, the latter on knowledge of the distribution of cells in the body. The epidemiological problem is to locate individuals with a standard pattern of behavior in the population at large or in special institutions; and the cytological problem is to locate cells of a given karyotype in particular regions of the body. It may be seen that the distribution of cases in the population is

Table 5. Brain Dysplasia and Dysfunction According to Karyotype

Aneuploidy	D Trisomy	E Trisomy	G Trisomy	Sources
Brain dysplasia	Holoprosencephaly cerebellar heterotopias	Arrhinencephaly; cerebellar, pesflocculi, vermis heterotopias; cerebral hemiatrophy; agenesis corpus collusum.	Small frontal lobes, cerebellum & brain stem; nodular swellings floccular peduncle; absent ganglia third cortical layer; senile plaque; Alzheimer's changes in youth.	Brun 1968; Crome 1967; Davidoff 1928; Finley et al. 1963; Gottlieb et al. 1962; Jelgersma 1963; Kakulas & Rosman 1965; Lewis 1964; Marin-Padilla et al. 1964; Norman 1966; Polani 1967; Terplan et al. 1970; Wertelecki et al. 1970
Neurological dysfunction	Multifocal epilepsy Hypertonia	Hypertonia	Hypotonia; delayed dissociation; motor automatisms, poor stereognosi.	
Psychological dysfunction	Severe mental retardation	Severe mental retardation	Low IQ	

Aneuploidy	XXX	XXY	XO	XYY XXYY	Sources
Brain dysplasia	Pituitary adenomata		Grey matter heterotopias; cerebellar medulloblastoma; cerebral gliomas; centrum ovale defects.		Alexander 1964; Kidd et al. 1963; Money 1963; 1964 & 1970; Money & Alexander 1966; Money & Mittenthal 1970; Shaffer 1962
Neurological dysfunction	Epilepsy	Essential tremor Epilepsy		Epilepsy	
Psychological dysfunction	Low IQ	Normal IQ distribution	Space form dysgnosia; emotional inertia.	Normal IQ	

Table 6. Epilepsy in Known Psycopaths with Previously Diagnosed Chromosomal Abnormality

Karyotype	Source	N	Epilepsy				References
			Clinical	EEG abnormal	EEG borderline	Overall %	
XYY	Penitentiary	11	–	0	5	40	Nielsen 1970
	Penitentiary	12	2	–	–	15	Daly 1969
XXY	Penitentiary	28	–	4	5	30	Nielsen 1970
	Mental institutions	71	5	–	–	7	Hambert 1964, 1966
	Hard-to-manage school boys	10	–	3	1	40	Annell 1970
XXX	Female mental institution	22	2	–	–	9	Kidd et al. 1963
	Penitentiaries						
Males	34%						
Females	–		Institutions 18%				
			9%				

Sporadic case reports (biased ascertainments: patients referred for clinical study & studied by EEG and karyotype simultaneously)

Karyotype	Source	N	Epilepsy		References
			Clinical	EEG	
XO	Female psychiatry clinic	4	4	4 (temporal & parietal lobe)	Mellbin 1965
XXY	Penitentiary	1		+	Welch et al. 1967
	Penitentiary	1		+	Wiener et al. 1968
	Psychiatry clinic	1		+	Kelly & Almy 1967
XXX	Female mental institution	1		+	Fraser et al. 1960

Table 7. Sex Chromosomal Anomalies in Previously Ascertained Epileptics

With mental illness	N	Karyotype	Frequency	References
Male epileptics	71	XYY	1 (1.2%)	Court Brown 1968
Female epileptics	100	XXX	3 (3%)	Hassing 1965
Sporadic case reports	1	XXY	–	Hambert & Frey 1964
	1	XYY	–	Rainer 1969
	1	XXX	–	Olanders 1967
Without mental illness				
Female epileptics	1,030		0	Akesson 1969

predicated on the distribution of cells in the body, and that methods of ascertainment and sampling, to be most useful, must take account of the likeliest times and points of connection between the two variables, social and biological. Cytogenetically speaking, the simplest choice is to study sexual abnormalities, because sex chromosomal anomalies are commoner than autosomal trisomies, perhaps because they are less often lethal and, more particularly, because they have reasonably well-specified anatomical and physiological correlates. In addition, among the varieties of dangerous or severely abnormal behavior, sexual disorders are easier to distinguish than, say, indiscriminate violence against persons and property, or antisocial activities directed against economic, political, or family circumstances. They are frequently stereotyped, often have a recognizable, immature pattern (i.e., they would be "normal" in childhood) and need not be accompanied by other types of criminal or psychotic behavior.

Prison inmates tend to come from classes poorly attended by the medical system; whereas hospital patients by definition are not lacking in this respect. It may be, therefore, that most of the criminal aneuploids would also be found to have evidence of physiological abnormality if their self-recognition was as high, or if they were examined with the same sophistication as cases attending hospitals, and that they are suffering from essentially the same disease even though the dangerousness and repetition of their offenses leads them involuntarily to jail instead of voluntarily to hospitals. In part, the difference between the two groups appears to be one of upbringing and education. For obvious reasons, all intersexuals in a sexually polarized society have some degree of sexual psychoneurosis or even psychosis; but only some, those who tend to be caught by the legal system, appear to solve their problems by excessive violence, perhaps because a certain amount of aggressiveness is necessary to deal with the limitations and stresses of life in their part of society. Experience with intersexual criminals suggests that the problem is not one of violence per se, but of a sexually immature emotional lability which goes along with a social readiness to use violence and an inescapable frequency of environmental stress that elicits fear and depression.

Differentiation

If the underlying emotional conflict is sexual, one might predict that sex chromosomal abnormalities would play a part in such diseases, provided that the sexual centers of the brain were preferentially affected by the genetical lesions. Beyond this, the distribution of abnormal cells through the nervous system may have led to additional involvement of brain circuits governing rage or attack (motor) behavior, or made them receptive to adjacent epileptiform discharges. Implicit in this theory is that the brain alone need be disturbed by cytogenetical displasia, and no other part of the body. The evidence for this is that intersexuals are often phenotypically quite normal, at least on routine medical examination of the outside of the body (this may be why they are missed medically, instead being "ascertained" by the legal system), and the cytological basis for this could be that they are mosaics and that the abnormal cells lie deep in the limbic system. Alternatively, even if the cellular changes are body-wide, they may have a deleterious effect only on those tissues in the neuroendocrine axis crucially dependent on sex chromosomal balance: genitalia, gonads, and sexual integrating centers of the limbic system. One can surmise that 100 percent aneuploidy of all tissues will result in sufficient disorganization of primary and secondary maturation in the tissues at risk to make it very likely that the patient will be seen in hospital; a diffuse mosaicism (all tissues equally, but less than fully, loaded) may affect body habitus and mental status to a lesser extent, the patient's progress depending upon tolerance of his environment and upbringing to his physical appearance and sexual psycho-type; and a patchy or multifocal distribution, varying from tissue to tissue, will only affect behavior if one such mosaic area exists in the limbic brain. In the extreme case, a single focus may occur in one site only: the ultimate expression is, of course, a tumor, the constituent cells of which become increasingly aneuploid as the tumor progresses toward malignancy. If, in any of these mosaics, the brain is not involved, behavioral disorder is less likely to be a primary symptom, the patient instead suffering from a variety of somatic defects known to occur with chromosomal abnormality, some of which may provoke a reactive neurosis, though not an aggressively disposed one.

Embryogenesis

The origin of these somatic variations depends upon the time at which the aneuploidy occurs: before conception (i.e., originating in parental gonads) in full aneuploids; and at various post-zygotic stages in the mosaics. Beyond

this the natural history of the disorder depends upon the migration and proliferation of aneuploid cells during differentiation and maturation.
Aneuploid cells tend to respond inaccurately to embryological induction
and internal selection (Razavi 1972), the result being a variety of abnor-
mally placed and poorly formed organs. The immediate result is a breakdown in humoral or neurophysiological circuitry, partly because the cells
are in the wrong place and partly because they are unable to perform
normal metabolic functions. This is easiest to see in the disturbance of sex
hormone and gonadotrophin circuitry (Table 4), and it also seems to be at
the root of electroencephalographic disturbances (Table 6).

To investigate this situation we need tests of variation in sexual
dimorphism in the limbic system and neuroendocrine axis. There are
several ways in which this might be approached, most usefully perhaps
in animal models; but before this is done it would be helpful to show the
existence of such variation in the tissues at risk, either directly or con-
tingently, in patients in whom epileptiform or limbic brain damage and
behavioral anomaly are consistently associated.

Impulsive Temporal Lobe Epileptics: Cytogenetics and Dermatoglyphics

One way to test this possibility—that sexual aneuploidy makes emotional
outbursts more likely because of a direct effect on the limbic system—is to
do chromosomal tests on individuals ascertained because they are known
both to have limbic brain disease and to be habitually aggressive. A suitable
group are temporal lobe epileptics, in whom the predominant symptom is
uncontrollable loss of temper. Twenty-two such cases were examined
cytogenetically with the results shown in Table 8.

The preliminary conclusions are that sex chromosomal aneuploidy is
more frequent among male temporal lobe epileptics, whose main com-
plaint is impulsive rage, than in the general population. These results
suggest a new nosological criterion for the disease, based on genetical
vulnerability. Apart from the increased frequency of genetical anomalies
(equivalent to the rate in sexual offenders), a striking feature of this group
of temporal lobe epileptics is that the lesions are predominantly on the left
side of the brain, and that many more males presented themselves for
diagnosis than females. All the males have left-sided lesions. No reliable
conclusions can be made about females because the sample size is too
small (the simplest explanation for this being that the disease affects
females less frequently than males), but right-sided lesions were slightly
commoner than left (Sherwin, I., personal communication).

If one combines the cytogenetical findings with neurological data on
unilateral predominance, a pattern emerges which supports Ounsted's
(1966) and Taylor's (1969) speculations on sex-dependent variation and differential cerebral hemispheric maturation in temporal lobe epilepsy.

Table 8. Sexual Aneuploidy with Agressive Behavior in Temporal Lobe Epileptics

Patients studied	Male	Female
Number	18	4
Aneuploid	2	0
Karyotype	XXY, XYY	
Overall rate	10%	
Sexual offenders	10%	
General population	.18%	.08%

Female skeletal and cerebral maturity is persistently in advance of males during early childhood, but female tissues stop growing at an earlier age than those in males, so that the final size of female organs is smaller than in males. Thus, XXY children grow at the same rate as XY (normal males); but XO's (Turner's syndrome) grow as fast as XX's (normal females)—i.e., faster and sooner completed than XY's. The differences appear to be due to a quantitative effect of sex chromatin: XXY's grow less extensively than XYY's. In parallel with these differences, Taylor showed that age-dependent febrile convulsions in infancy are followed by temporal lobe epilepsy more often in boys and most often in the larger-sized (longest-growing) hemisphere of either sex. In both sexes cerebral convulsions, anteceding epilepsy, occur between the ages of six months and five years, with a maximum incidence at eighteen months. In boys, the incidence of cerebral convulsions falls steadily and smoothly between eighteen months and four years, while in girls the fall is sudden, most attacks having occurred by two years of age. Subsequently, temporal lobe epilepsy occurs most frequently in the dominant (usually larger, left) lobe of males, and least often in the nondominant (usually right) lobe of females. The implication is that the larger hemisphere of males is at hazard for the largest period and the smaller hemisphere in females for the shortest duration, precisely because maturation takes longer in males and larger structures of either sex. Hence, the increased male:female ratio in temporal lobe epilepsy and hence also, perhaps, the increased frequency of males exhibiting severe behavior disorder with left-sided lesions.

Two growth defects are discernible, one dependent on genetic sex, the other on laterality. On this basis one might predict that cases with excess chromosomes, and most particularly those containing one or more Y chromosomes, will be at maximum hazard to epileptogenic insults, because of the longer period of cerebral maturation which they undergo. Table 5 bears this out: if behaviorally ill intersexes already known to carry chromosome abnormalities are searched for epilepsy, the rates vary between 7 and 40 percent, the highest frequency occurring in male prison inmates with increased sex chromosomes, whether XXY or XYY, and the lowest in females at (noncriminal) mental institutions and in XO's. If the reverse procedure is carried out—studying known epileptics for chromosomal anomaly—the rates are lower, but still above normal in mentally-ill epileptics of either sex; whereas they are normal or even below normal in female epileptics without mental illness (Table 7). This suggests that the onset of epileptogenic disorders in the mentally ill may be the earliest warning of a general genetic tendency to progressive neural deterioration, and that this in turn depends upon prolonged delay in maturation, during a period of time when aneuploid cells are attempting to migrate and multiply relatively slowly and inaccurately. The important point to note is that mental defectives, in whom neurological deterioration is general and severe, presumably die earlier or are not studied as intensively as the mentally ill, in whom fewer cells or less extensive areas of tissue are affected and who are often studied at hospitals rather than being directly institutionalized early in life. In either event, the early behavioral liability of mentally-ill epileptics with chromosomal abnormality who are yet to deteriorate intellectually may be more likely to exist in open society where social consequences are broader than in institutions. The sporadic case reports listed in Tables 6 and 7 show that chromosomally abnormal epileptics with mental illness are not hard to find in the literature: presumably this is because they are not hard to find in psychiatric and neurological clinics aware of the conjunction. Care must be taken with clinical reports, however, because ascertainment is via electroencephalographic abnormality and may be over-readily diagnosed. In the series reported here, in which ascertainment was based on the presence of violence as well as an epileptiform dysfunction, there was a heavy predominance of males and of left-sided lesions, together with the occurrence of two cases with increased sex chromosomal abnormality (one with an extra X and the other with an extra Y). This is most simply explained on the bases of *prolonged* maturation, which laid neural tissues open to epileptogenic insult, and *delayed* maturation, which resulted in dysmature integration of limbic system function. It seems possible that further studies may uncover cases with mosaic anomalies in whom selective damage occurs in the nervous system, and in whom the behavioral disorders are severe, in spite of the presence of an apparently small proportion of abnormal cells elsewhere in the body.

No direct conclusions can be drawn about the relationship of the chromosomal aneuploidies, diagnosed in the blood of these patients, to the tissue damage apparently at the origin of the temporal lobe epilepsy and its behavioral manifestations. Some contingent information on developmental malformation in the ectoderm comes from the observation that many of these cases had abnormal dermatoglyphics, whether or not chromosomal changes were detected. Skin and brain both derive from ectoderm, and some of the abnormalities in fingerprints appear to be

Table 9. Dermatoglyphic T-Tests: Temporal Lobe Epileptics vs. Normals

| Laterality of temporal lobe lesions detected | Digital | | | | | |
| | Pattern types | | | Ridge counts | | |
	Both hands	R.h.	L.h.	Both hands	R.h.	L.h.
Male: (left>>>right) (26)	0.005*	0.10	0.005	0.02*	0.02	0.01
Female: (right>left) (11)	0.005	0.005	0.01	0.20	0.20	0.50

*Mainly small ulnar loops.

related to alterations of normal sexual dimorphism in this germ layer. Chief among these is a significantly lowered ridge count with a concomitant decrease in the size of patterns, changes typically associated with increases in sex chromosomes, and a tendency toward similarity of patterns, chiefly small ulnar loops as opposed to the normal mix of large and small loops, whorls, and arches prevalent in the general population. Other abnormalities appear to be related to changes in body symmetry and growth patterns. The distribution of dermatoglyphic variation across the two sides of the body is anomalous, the most striking characteristic being a unilateral change in ridge counts and pattern frequencies (Tables 9, 10) according to the side of the lesion. Thus, in males with a left-sided neurological lesion, the largest part of the variance is due to the left hand; while in female patients, where right lesions are slightly more frequent than left, there is an equivalent contribution of the right hand to the overall difference from normal females. This suggests that there is a general disturbance of neuroectodermal symmetry about the midline and that, in some instances, dermatoglyphics may assist in confirming the side of a developmental lesion. They may also elucidate the original focus in bilateral disease. In a series of temporal lobe epileptics (Falconer, personal communication) hamartomata were found in 22 percent of cases undergoing unilateral lobectomy. These patients were characterized by two important features: the disease tended to start in childhood, and

Table 10. Dermatoglyphic T-Tests: Temporal Lobe Epileptics vs. Normals

| | Palmar | | | | | |
| | ATD angle | | | A-B ridge count | | |
	Both hands	R.h.	L.h.	Both hands	R.h.	L.h.
Males (26)	0.90	0.90	0.90	0.40	0.40	0.40
Females (11)	0.05	0.20	0.20	0.40	0.20	0.90

whether or not the lobectomy was performed early, there was a high frequency of subsequent schizophrenia in late adolescence or early adult life. Therapeutic response was not as good as in patients suffering from mesial temporal sclerosis. Schizophrenia has strong genetic determinants and is not uncommonly associated with abnormal dermatoglyphics, while a harmatomatous anomaly arises from a developmental alteration in the brain. All this implies that if there is developmental limbic damage in mentally-ill epileptics, the use of neuroectodermal markers, such as dermatoglyphics, should be assessed in detection of liability early in life.

Animal Models

Studies on gynandromorphic Drosophila behavioral mutants by Hotta and Benzer (1970) and Ikeda and Kaplan (1970) have much to say about these behavioral illnesses in human sexual aneuploids. In Drosophila normal males are XO and females XX, a partial version of the sex chromosomal variation which can occur in humans. Individual fruit-flies may be produced in which one sector of the body, the eye or the local ganglia controlling leg movement, is male (XO) and the rest of the body is female (XX). To the untutored eye, these mosaics are of largely normal phenotype, precisely because small segments of the body are affected: the situation is closely analogous to human sexual mosaics unexpectedly discovered in penitentiaries, instead of hospitals. Hotta and Benzer (1972) provide evidence for independent function of these patches of intersexual nerve cells in Drosophila. If the male sector carries an X-linked recessive mutant (in the eye, a failure to respond to phototactic stimuli; in the limbs, an abnormal shaking after etherization), the affected area displays the mutation regardless of the dominant allele present in the corresponding female tissue. That is, there is failure of neurophysiological integration between sexually mosaic proportions of the neural system, and this blocks the ability of the normal allele to compensate for the recessive and interferes with cooperation between the two sides of the body. No "emotions" are of course detectable in Drosophila gynandromorphs, except insofar as the mutants exhibit a primitive version of neurotic indecision in the completion of an instinctive pattern of acts, the most interesting available for analysis being courtship and mating patterns. But the motor components of the mosaic disturbance in the animal's neurophysiology are beautifully clear. In humans, there is a motor element of the emotional illness—attack behavior—which seems to be the one problem which the patient largely denies ability to control, or even recognize, though "he" may be very willing to recognize the underlying emotional conflict and pleads for help precisely for the sexual or depressive psychoneurosis. Very careful investigation, however, can uncover epileptiform dysfunction in the limbic (emotional) brain, usually on one side and sometimes associated with small developmental anomalies or hamartomata which have failed to integrate anatomically (or physiologically) with surrounding tissues (Corsellie 1970). The results are most often seen as defects of function in the "dominant" side of the brain: no contralateral compensatory re-integration occurs. Without drawing false analogies between functional and genetical dominance, it is interesting to speculate that the failure of cross-compensation between genetically dissimilar cells on either side of the brain may be at the root of the (epileptiform) impulsive behavioral outbursts so characteristic of these individuals. These defects could occur in human sexual mosaics of the types XO/XX, XO/XXX, XO/XXY, etc. Recessive sex-linked genes may display their deleterious effects in XO-sectors of the body. In the brain lack of compensatory integration would lead to focal disorders characterized by release phenomena and lack of control in behavioral patterns, or circumscribed cognitive defects of the sort frequently described in Turner's syndrome (Money 1970). The defect here is in electrophysiological integration between genetically dissimilar cells, some of which may also be poorly positioned for neurohumoral integration and suffer a double jeopardy.

Conclusions

Clinically, it seems that temporal lobe epileptics, in whom behavioral symptoms are predominant, deserve cytogenetic diagnosis as part of routine management, and that their families, too, should be investigated, because combinations of this sort tend to recur in pedigrees. The diagnosis may have prophylactic and therapeutic significance:

1. In some patients there may be a higher likelihood of developmental temporal lobe tumors (hamartomata) requiring temporal lobe lobectomy. Where there is a suspicion that the patient has a space-occupying lesion, on the basis of neurological studies, and he is known to have chromosomal abnormality, the two diagnostic features taken together may be regarded as additive indications of the presence of a tumor.
2. Knowledge of the karyotype may provide a basis for psychodynamic analysis, and its correlation with hormone balance may be useful for endocrine therapy related to genetic output.
3. Several somatic complications of chromosomal aneuploidy, unrelated to behavioral disorder (leukemia, gonadal cancer, fertility problems, diabetes, autoimmunization) are serious enough to warrant early detection for medical reasons alone, and these patients should there-fore be kept under clinical surveillance. There is also the possibility that dermatoglyphics may prove a clinically useful route to early prediction of

epileptogenesis and protection from febrile convulsions, which appear to be at the origin of temporal lobe epilepsy in some pedigrees but not others.

In research, new tests are needed for the detection of ectopic foci resulting from abnormal migration of aneuploid cells; and the presence of odd molecular species of hormones produced by proliferation of genetically unbalanced glandular tissues. Several genetic lesions are now identifiable involving humoral circuitry between endocrine tissues producing differing (mutable) molecular species of sex hormones, especially sex hormones, and end-organs with (similarly mutable) receptors and metabolic pathways (Weiss & Lloyd 1972). It may be possible, by taking advantage of neural biopsies available from patients undergoing temporal lobectomy, to study the differential sex chromosomal content of constituent cells of these areas. The fundamental questions, biologically, are developmental and concern the mechanisms which govern migration and multiplication of mixoploid cells, diffusely or in foci, through germ layers or secretory anlage and their target organs. Some questions worth answering are:

1. What is the *in vivo* evolution of karyotype (XXY - XO/XY, etc.) and how does somatic transmission differ from germ cell transmission?
2. What governs the selective distribution of cells in the body according to demands of internal and external environments?
3. How are normal and abnormal metabolites integrated by the neuro-endocrine axis, when circuitry is disturbed by malplacement and malfunction of its constituent tissue masses?
4. How far can the components of behavior and their underlying physiological substrates be isolated and given sufficiently discrete definition to allow their use in ascertainment?
5. How far can dermatoglyphics be used as a measure of lateralization and focalization in ectodermal morphogenesis and epigenesis and, in particular, how may these be referred to dominance of symptomatology, or variation in electroencephalographic findings?

In institutional work it seems equally desirable to provide cytogenetic diagnosis for all offenders presenting with serious and repetitive behavioral disorders, especially if these are accompanied by psychoses, provided proper safeguards can be taken to prevent misuse of this information by legal authorities and as long as no attempts are made to use the information as a basis for physical therapy in jail, with the possible exception of unequivocally diagnosed brain tumors. Instead, advantage should be taken of the information for the patient and his psychotherapist to elucidate the sexual components of "his/her" personality, which by its very nature runs counter to the standard dictates of conventional society. In this way it may be possible to validate the patient's complaints in terms of real social difficulties, which arise from sexual ambiguities and lead to behavioral isolation. It should be possible to manipulate or select social environments so that patients are able to develop patterns of sexual behavior suitable to their own make-up without causing pain to others. None of this need involve the use of physical or drug therapy, most importantly because the genetical diagnosis is one of risk rather than of the actual presence of treatable physical disorder. In handling these patients it seems most rational to integrate social background with the natural history of the patient's disorder in society at large, and thereby select a more precise training and education for each individual according to psychic and biological character. Present social experiences may have additive or even multiplicative effects upon the progress of psychosexual neurosis, and the chief value of a genetical diagnosis is that it allows a taxonomical classification of sexual type, at an early stage in a troubled life, on which one may base the development of the patient personality according to social and educational opportunities. It seems fair to say, therefore, that cytogenetical diagnoses should be built into the routine medical examination of selected classes of offender, and that the epidemiological data so far collected provide good justification for allocation of mental and public health resources in prison health, a field until recently neglected, but which is now recognized to have the same importance as other areas of chronic mental ill-health. In this way, it may be possible to extend medicine and psychiatry into a new field, in much the same way as the early recognition that madness was due to mental illness and not satanic possession led to the humane treatment of the mentally ill.

References

Akesson, H. O., & Olanders, S. Frequency of negative sex chromatin among women in mental hospitals. *Human heredity*, 1969, **14**, 43–47.

Alexander, D., Walker, H. T., & Money, J. Studies in direction sense (1. Turner's Syndrome). *Archives of General Psychiatry*, 1964, **10**, 337–339.

Annell, A. L., Gustavson, K. H., & Tenstam, J. Symptomatology in schoolboys with positive sex chromatin (The Klinefelter Syndrome). *Acta Psychiatrica Scandinavica*, 1970, **46**, 71–80.

Brun, A., & Skold, G. CNS malformation in Turner's Syndrome (an integral part of the syndrome?). *Acta Neurophathologica*, 1968, **10**, 159–161.

Casey, M. D., Blank, C. E., Street, D.R.K., Segall L. J., McDougall, J. H., McCraft, P. J., & Skinner, J. C. XYY chromosomes and antisocial behavior. *Lancet*, 1966, **2**, 859–860.(b)

Casey, M. D., Segall, L. J., Street, D.R.K., & Blank, C. E. Sex chromosome abnormalities in two state hospitals for patients requiring special security. *Nature*, 1966, **209**, 641–642. (a)

Corsellis, J.A.N. The pathological anatomy of the temporal lobe with special reference to the limbic areas. In Price (Ed.), *Modern trends in psychological medicine*, Vol. 2. New York: Appleton-Century-Crofts, 1970, pp. 296–325.

Court Brown, W. M. Males with an XYY sex chromosome complement. *Journal of Medical Genetics*, 1968, **5**, 341–359.

Crome, L., & Stern, J. *The pathology of mental retardation.* London: Churchill Livingstone, 1972.

Daly, F. Neurological abnormalities in XYY males. *Nature*, 1969, **221**, 472–473.

Davidoff, L. M. The brain in mongolian idiocy. *Archives of Neurology and Psychiatry*, 1928, **20**, 1229–1257.

Finley, W. H., Finley, S. C., & Carte, E. T. 17–18 trisomy syndrome. *American Journal of Diseases of Children*, 1963, **106**, 591–596.

Forssman, H., & Hambert, G. Incidence of Klinefelter's Syndrome among mental patients. *Lancet*, 1963, **1**, 1327.

Fraser, J. H., Campbell, J., Macgillivray, R. C., Boyd, E., & Lennox, B. The XXX Syndrome frequency among mental defectives and fertility. *Lancet*, 1960, **2**, 626–627.

Goodman, R. M., Smith, W. S., & Migeon, C. J. Sex chromosome abnormalities. *Nature*, 1967, **216**, 942–943.

Gottlieb, M. I., Hirschhorn, K., Cooper, H. L., Lusskin, N., Moloshak, R. E., & Hodes, H. L. Trisomy-17 syndrome. *American Journal of Medicine*, 1964, **33**, 763–773.

Hambert, G. Positive sex chromatin in men with epilepsy. *Acta Medica Scandinavica*, 1964, **175**, 663–665.

Hambert, G. *Males with positive sex chromatin.* Goteborg: Academiforlaget-Gumperts, 1966.

Hambert, G., & Frey, T. S. The electroencephalogram in Klinefelter Syndrome. *Acta Psychiatrica Scandinavica*, 1964, **40**, 28–36.

Hassing, F. Cited by Forssman, H. The mental implications of sex chromosome aberrations. *British Journal of Psychiatry*, 1970, **117**, 353–363.

Hotta, Y., & Benzer, S. Genetic dissection of the Drosophila nervous system by means of mosaics. *Proceedings of the National Academy of Sciences* (Washington, D.C.), 1970, **67**, 1156–1163.

Hotta, Y., & Benzer, S. Mapping of behaviour in Drosophila mosaics. *Nature*, 1972, **240**, 527–535.

Hunter, H. Chromatin-positive and XYY boys in approved schools. *Lancet*, 1968, **1**, 816.

Ikeda, K., & Kaplan, W. D. Unilaterally patterned neural activity of gynandromorphs, mosaic for a neurological mutant of Drosophila melanogaster. *Proceedings of the National Academy of Sciences* (Washington, D.C.), 1970, **67**, 1480–1487.

Ismail, A.A.A., Harkness, R. A., Kirkham, K. E., Loraine, J. A., Whatmore, P. B., & Brittain, R. P. Effect of abnormal sex chromosome complements on urinary testosterone levels. *Lancet*, 1968, **1**, 220–222.

Jacobs, P. A., Brunton, M., Melville, M. M., Brittain, R. P., & McClemont, W. F. Aggressive behaviour, mental sub-normality and the XYY male. *Nature*, 1965, **208**, 1351–1352.

Jelgersma, H. On the tuber flocculi in mongolian idiocy. *Psychiatria, Neurologia, Neurochirurgia* (Amsterdam), 1963, **66**, 131–137.

Kakulas, B. A., & Rosman, N. P. 13–15 trisomy in eight cases of arrhinencephaly. *Lancet*, 1965, **2**, 717–718.

Kelly, S., & Almy, R. Another XYY phenotype. *Nature*, 1967, **215**, 405.

Kidd, C. B., Knox, R. S., & Mantle, D. J. A psychiatric investigation of triple-X chromosome females. *British Journal of Psychiatry*, 1963, **109**, 90–94.

Lewis, A. J. The pathology of 18 trisomy. *Journal of Pediatrics*, 1964, **65**, 92–101.

Marin-Padilla, M., Hoefnagel, D., & Benirschke, K. Anatomic and histopathologic study of two cases of D (13–15) trisomy. *Cytogenetics*, 1964, **3**, 258–284.

Mellbin, G. Neuropsychiatric disorders in sex chromatin negative women. *British Journal of Psychiatry*, 1965, **112**, 145–148.

Melnyck, J., Derencsenyi, A., Vanasek, F., Rucci, A. J., & Thompson, H. XYY survey in an institution for sex offenders and the mentally ill. *Nature*, 1969, **224**, 369–370.

Money, J. Cytogenetic and psychosexual incongruities with a note on space-form blindness. *American Journal of Psychiatry*, 1963, **119**, 820–827.

Money, J. Two cytogenetic syndromes: psychologic comparisons: intelligence and specific-factor quotients. *Journal of Psychiatric Research*, 1964, **2**, 223–231.

Money, J. Behavior genetics: principles, methods and examples from XO, XXY and XYY syndromes. *Seminars in Psychiatry*, 1970, **2**, 11–29.

Money, J., & Alexander, D. Turner's Syndrome: further demonstration of the presence of specific cognitional deficiencies. *Journal of Medical Genetics*, 1966, **3**, 47–48.

Money, J., & Mittenthal, S. Lack of personality pathology in Turner's Syndrome: relation to cytogenetics, hormones and physique. *Behavior Genetics*, 1970, **1**, 43–56.

Nielsen, J. Criminality among patients with Klinefelter's Syndrome and the XYY Syndrome. *British Journal of Psychiatry*, 1970, **117**, 365–369.

Norman, R. M. Neuropathological findings in trisomies 13–15 and 17–18 with special reference to the cerebellum. *Developmental Medicine and Child Neurology*, 1966, **8**, 170–177.

Olanders, S. Double Barr bodies in women in mental hospitals. *British Journal of Psychiatry*, 1967, **113**, 1097–1099.

Ounsted, C. *Biological factors in temporal lobe epilepsy*. London: Heinemann Medical Books, 1966.

Polani, P. E. Chromosome anomalies and the brain. *Guys Hospital Reports*, 1967, **116**, 365–396.

Rainer, J. D., Jarvik, L. F., Abdullah, S., & Kato, T. XYY karyotype in monozygotic twins. *Lancet*, 1969, **2**, 60.

Razavi, L. Sex chromosomal anomalies in sexual offenders. Paper delivered at the AAAS annual meeting, Symposium on Biology and Sociology of Violence, Boston, 1969.

Razavi, L. Rate of chromosomal change in criminal populations. Paper delivered at the symposium on behavioral genetics, A.I.B.S., Detroit, 1970.

Razavi, L. Cytogenetic and somatic variation in the neurobiology of aggression: epidemiological, clinical and morphogenetic considerations. In W. S. Fields & W. H. Sweet (Eds.), *Neural bases for violence and aggression*. St. Louis: Warren H. Green, 1973, in press, June, 1974.

Santen, R. J., deKretser, D. M., Paulsen, C. A., & Vorhess, J. Gonadotrophins and testosterone in the XYY Syndrome. *Lancet*, 1970, **2**, 371.

Shaffer, J. W. A specific cognitive deficit observed in gonadalaplasis (Turner's Syndrome). *Journal of Clinical Psychology*, 1962, **18**, 403.

Shapiro, L. R. Hormones and the XYY Syndrome. *Lancet*, 1970, **1**, 623.

Taylor, D. C. Differential rates of cerebral maturation between sexes and between hemispheres (evidence from epilepsy). *Lancet*, 1969, **2**, 140–142.

Telfer, M. A., Baker D., Clark, G. R., & Richardson, C. E. Incidence of gross chromosomal errors among tall criminal American males. *Science*, 1968, **159**, 1249–1250.

Terplan, K. L., Lopez, E. C., & Robinson, H. B. Histologic structural anomalies in the brain in trisomy 18 syndrome. *American Journal of Diseases of Children*, 1970, **119**, 228–235.

Wegmann, T. G., & Smith, T. W. Incidence of Klinefelter's Syndrome among juvenile delinquents and felons. *Lancet*, 1963, **1**, 274.

Weiss, J., & Lloyd, C. Hormones and violence. In W. S. Fields & W. H. Sweet (Eds.), *Neural bases for violence and aggression*. St. Louis: Warren H. Green, 1974, in press.

Welch, J. P., Borgaonkar, D. S., & Herr, H. M. Psychopathy, mental deficiency, aggressiveness and the XYY Syndrome. *Nature*, 1967, **214**, 500–501.

Wertelecki, W., Fraumeni, J. F., & Mulvihill, J. J. Nongonadal neoplasia in Turner's Syndrome. *Cancer*, 1970, **26**, 485–488.

Wiener, S., Sutherland, G., Bartholomew, A. A., & Hudson, B. XYY males in Melbourne Prison. *Lancet*, 1968, **1**, 150.

6 AN ADOPTIVE STUDY OF PSYCHOPATHY: PRELIMINARY RESULTS FROM ARREST RECORDS AND PSYCHIATRIC HOSPITAL RECORDS

Raymond R. Crowe

Introduction

The adoption method as a means of studying the genetics of human behavior was first used in the 1930's by Skeels (1938) to study the inheritance
of intelligence. Roe (1944–45) used this method to study alcoholism in
the early 1940's. However, it was not until its success in studying the
genetics of schizophrenia in the 1960's that it attracted wide psychiatric
attention. The adoption method can actually be broken down into two different methodologies. Rosenthal (1970, pp. 126–29) termed these the
"adoptees study method" and the "adoptees' families method." The latter
entails the study of the biologic relatives of an adoptee who possesses
the disorder under study. Dr. Hutchings' paper in this volume deals with
such a study of criminality. The former method employs a follow-up of
children born to a parent possessing the disorder under study and adopted
in infancy. This report will describe the results to date of an adoptees
study of psychopathy.

The first prerequisite for an adoptees study is a large group of adults
who have the disorder being studied and who are likely to have given up
children for adoption. Further, records must be available to search for
probands. This limits the study to disorders which are likely to institu-
tionalize a parent, especially a mother, and which at the same time are
likely to cause a broken home. Psychopathy fits these criteria and, therefore, would be a logical disorder to study.

Psychopathy would be a logical choice for other reasons as well. First,
the personality disorders have not been as well studied genetically as the

Raymond R. Crowe, M.D. Department of Psychiatry, University of Iowa, College of
Medicine, Iowa City, Iowa.

functional psychoses, and psychopathy being one of the more severe of these would be a logical starting place. Second, the studies of criminal twins have all shown higher concordance between monozygotic than between dizygotic twins. Especially pertinent are the recent studies by Yoshimasu (1961) and Christiasen (1970) which are methodologically well done and both of which show a significantly higher concordance between monozygotic than between dizygotic twins. Finally, Schulsinger (1972) has conducted an adoptees' families study of psychopathy which has demonstrated that genetics are important in the development of the disorder.

This paper describes the present status of an adoptees study of psychopathy. The study has entailed a follow-up into adult life of a group of probands born to incarcerated women offenders. The assumption is made that psychopathy would be prevalent among such a group of women offenders. This assumption is supported by Cloninger's finding of "sociopathy" in 65 percent of a group of convicted women felons (Cloninger & Guze 1970). The follow-up has consisted of obtaining all possible records on the adopted offspring, plus personal interviews with each possible subject eighteen years or older. A previous report (Crowe 1972) describes those probands ascertained through arrest records in the state in which the subjects were reared. This paper focuses on those subjects plus subjects ascertained through psychiatric hospital records in the light of more recent data obtained from the interview follow-up.

Methods

Over the years 1925 to 1956, a group of 52 probands was obtained—27 males and 25 females. They were born to 41 female offenders, 38 from the State Women's Reformatory and 3 from the State Training School for Girls who had subsequently committed an adult offense. Cases were obtained by searching records at the institutions and by cross-checking names of women admitted to the reformatory against index files of adoptions in the state offices. The probands' ages at follow-up ranged from 15 to 46 years with a mean of 25.6 years. All were white.

The only information available on the biologic parents of the proband group was obtained from the prison records on the mothers. Ninety percent of these mothers were felons. The most frequent offenses were check felonies, with 15 women sentenced, prostitution 5, larceny 4, desertion 3, adultery 3, breaking and entering 2, lewdness 2, and one each assault, conspiracy, aiding prisoners to escape, bigamy, transmitting a venereal disease, contributing to the delinquency of a minor, and exposing a dead body. This last offense represented the abandonment of a baby and the charge was reduced from murder to exposing a dead body. No information is available on the biologic fathers of most of the probands.

A group of adopted controls was selected from the state index of adoptions by selecting the nearest entry to the proband's which matched for age, sex, race, and approximate age at the time of the adoptive decree. An age match was difficult on subjects born prior to 1945 due to a scarcity of cards, so they were matched within five years of the proband's date of birth. This gave a control group with a mean age of 24.6 years, one year less than that of the proband group. Due to the brief records, no information was available on the biologic parents of the control group.

Table 1 represents an attempt to evaluate the adequacy of the control group in controlling for those variables which are important in a study of psychopathy. In order to assess the effect of parental deprivation, the age at maternal separation, age at adoptive placement, and length of time spent in orphanages were examined. The larger adoption agencies around the state were very helpful in providing this information on cases they had handled. In addition, subjects who admitted to being adopted during the interview were asked their age at the time of adoption. Unfortunately this data could not be retrieved on every subject, but data on age at separation, age at placement, and thus length of time spent in orphanages or foster homes, was obtained on 70 subjects: 38 probands and 32 controls. The probands were separated from their natural mothers at a mean age of 3.9 months (range 0–18 months) and placed at a mean age of 9.5 months (range 0–26 months). The controls were separated from their mothers at a mean age of 3.7 months (range 0–48 months) and placed at a mean of 7.1 months (range 0–48 months). Neither difference approached significance by chi square analysis. The third line shows the time spent in orphanages or foster homes prior to placement. The probands spent a mean of 5.6 months (range 0–24) and the controls spent a mean of 3.4 months (range 0–24). This difference likewise was not significant.

Data on socioeconomic status of the adoptive parents was based on the father's occupation ascertained by the interview. This was rated on the seven-point occupational scale from Hollingshead and Redlich (1958,

Table 1. Adoptive Environment

	N[a]	Probands	Controls
Mean age at maternal separation in months	70	3.9	3.7
Mean age at adoptive placement in months	70	9.5	7.1
Mean time spent in orphanage in months	70	5.6	3.4
Socioeconomic status of adoptive homes	75	4.0	3.9
Broken homes	75	7	9
Divorce	75	2	2
Parental psychopathology[b]	75	7	4

[a]N=number of Ss, out of 104, on which data was obtained.
[b]This term represents personality disorders, drug addiction, and alcoholism (including heavy drinking).

pp. 390–91). This information was available on 72 percent of the subjects. The mean socioeconomic status of the probands' adoptive homes was 4.0 and that of the controls was 3.9. The interviews also provided information on the home environment for 72 percent of the overall group. Seven probands were reared in broken homes and two of these were broken by divorce. This compares with nine homes of controls which were broken and two of these were also broken by divorce. Finally, severe personality disorder, alcoholism, or drug addiction is often found in the parents of antisocials. In Table 1 these are grouped together under "parental psychopathology," which was found in seven homes in the proband group and four in the control group. From the above subsample of the overall group of subjects, it appears that the control group has provided an adequate control for the probands.

The data of this report have been obtained from arrest records from the State Bureau of Criminal Investigation, hospital records from the 4 state hospitals plus the state teaching hospital, social histories from various institutions, and finally personal interviews whenever possible with those subjects turned up by the previous record searches. At the time of this follow-up, 84 subjects (42 probands and 42 controls) had reached the age of 15 or older in the State of Iowa and were, therefore, considered to be at risk. Seventy-four were adults, 37 probands and 37 controls.

Results

The results from the arrest records are shown in Table 2. First, taking the adult records, 7 subjects had records and all 7 subjects had at least one conviction. Four had 2 or more arrests but only 2 had multiple convictions.

Table 2. Arrest Records

	Probands	Controls
Number of subjects at risk	37	37
Subjects with adult arrest	7	2 p^a=0.076
adult conviction	7	1 p=0.028
multiple arrests	4	0
multiple convictions	2	0
felons	3	0
Incarcerations		
Juvenile[b]	3	0
Adult	4	0
Total subjects incarcerated (N=84)	6	0 p=0.013

[a]Fisher's Exact.
[b]One juvenile was ascertained through hospital records and sent to the training school shortly afterward.

Three of these 7 probands are felons. This contrasted with the controls among whom only 2 had adult records and only one of these had a conviction. None of the controls were felons nor did any have records of multiple offenses. Juvenile court records were not available but it was possible to obtain accurate records on any individuals admitted to the State Training Schools as juveniles. There were 3 of these among the probands compared to none of the controls. Putting this together with the adult incarcerations, there have been 6 probands who have spent time in correctional institutions, compared to none of the controls.

Five of the probands with records are male and 4 are female. One male was sent to the training school as a juvenile for delinquency but was not at risk for an adult record as he had just turned eighteen. This subject was ascertained through his hospital records, because he was not sent to the training school until after arrest records had been checked. Another male was convicted of lascivious acts and served ninety days in the county jail. A third male was convicted of forgery and given one year probation. The fourth male was convicted twice of larceny and served eighteen months in the Men's Reformatory. The last male was convicted of obtaining money under false pretenses and placed on probation. In addition, he had a record of 6 misdemeanors. One of the females was committed to the Girls' Training School for delinquency and has no adult record. Of the remaining 3, one served thirty days in the city jail for petty larceny. Another was convicted and fined for a check misdemeanor. The third had been in the training school for delinquency as a juvenile, plus a conviction for lewdness as an adult, and a charge of embezzelment of an auto which was dismissed.

The psychiatric hospital records are shown in Table 3. Eight probands have been seen at the state hospitals and 7 have been hospitalized. Seven of the 8 probands were seen for antisocial behavior and 6 of these 7 were court-ordered evaluations. Of these probands, 6 have arrest records.* Thus, there is considerable overlap between the subjects with arrest records and those with hospital records. Two of the controls have been seen at one of the state hospitals and one of these was hospitalized. Both controls were seen for behavior disorders.

Combining the arrest records and hospital records, a total of 11 probands and 4 controls has been ascertained by searches of records in the state in which the subjects were reared. In order to determine if the hospital and arrest records are a valid indicator of psychopathy in the 2 groups, all of the material collected on cases was dictated into case histories and given to Dr. Paul Huston for blind diagnosis. This material consisted of structured interviews with 10 of the subjects, and interviews with the parents of an

---

*Another had a juvenile court record and was one of the court-ordered examinations, but this did not appear in the arrest records and he is counted as a hospitalization only.

Table 3. Psychiatric Hospital Records

		Probands	Controls
Number of subjects at risk		42	42
Subjects hospitalized		7	1 p*=0.03
Outpatient only		1	1
	Total	8	2 p=0.04

*Fisher's Exact.

additional 2 subjects. On the remaining 3 information was obtained from psychiatric and social histories from the various institutions. The results are shown in Table 4. Of the 11 probands ascertained, 6 were diagnosed antisocial personality. All 6 had arrest records and 5 of the 6 had also been picked up through psychiatric hospital records. Therefore, it was this group of probands which accounted for the majority of the overlap between the 2 sources of records. Of the remaining probands, 2 were diagnosed inadequate personality, 2 passive aggressive personality, and one was diagnosed schizoid personality. None of the controls were diagnosed antisocial personality. One was diagnosed passive aggressive personality, one hysterical personality, one was diagnosed homosexuality and/or transvestism, and the fourth received no psychiatric diagnosis. Therefore, the arrest and hospital records do indicate a high rate of antisocial personality among the probands, but this was not the case among the controls.

Discussion

In a previous publication (Crowe 1972), the arrest records were reviewed and it was concluded that the proband group had shown a significantly higher rate of antisocial behavior leading to arrest than had the control group. In addition, the records seem to indicate that the antisocial acts of the probands were, on the whole, of a more serious nature than those of the

Table 4. Diagnoses of Subjects

	Probands (N=42)	Controls (N=42)
Total subjects ascertained	11	4
Antisocial personality	6	0
Inadequate personality	2	0
Passive-aggressive personality	2	1
Schizoid personality	1	0
Hysterical personality	0	1
Homosexuality/transvestism	0	1
No mental disorder	0	1

controls. The present findings reinforce those of the previous report. Six of the probands not only had arrest records but had received psychiatric evaluation as a result of their behavior and 5 of these were court ordered. Antisocial behavior alone would not prove an increased prevalence of antisocial personality among the probands. However, a blind evaluation of these records has demonstrated that the preponderance of the antisocial behavior is due to 6 probands who were diagnosed as antisocial personalities. Although the investigation to date has turned up a large number of antisocial personalities among the probands and not among the controls, it would be incorrect to conclude on the basis of this data that the disorder is significantly more common among the proband group. These diagnoses were made on a select subsample of individuals who were ascertained through records in the state in which they were reared, and any final conclusion will have to be drawn from diagnoses of the entire group of subjects.

The mothers of the 6 antisocial probands had been sentenced for assault with intent to injure, desertion, contributing to the delinquency of a minor, prostitution, and 2 were sentenced for larceny. The proband born to the mother who deserted her family was placed in the orphanange at less than one month of age, and the desertion should not have been a factor in his outcome. Five of these mothers were felons and the sixth had been sent to the training school as a juvenile. In addition, 5 of the 6 mothers had records of repeated offenses. This data suggests severe personality disorders, presumably antisocial personality, on the part of the mothers. At the present time, no information is available on the biologic fathers of those probands who subsequently developed antisocial personalities. This would be important information to have because if heredity does contribute to the development of antisocial personality then those individuals with 2 antisocial parents are at an increased risk for that disorder. Robins (1966, p. 105) found that persons diagnosed "sociopathic personality," especially women, tended to marry spouses with serious behavior problems, such as excessive drinking, arrests, unfaithfulness, cruelty, and failure to support. It is likely that many of the biological fathers of the probands had the same personality disorders as Robins found and, therefore, the chance that a number of the probands had 2 antisocial biological parents is real. In the future, every effort will be made to learn as much as possible about the natural parents of the probands.

The proband group was compared with the control group on the important environmental variables which have frequently been found in the early life histories of persons with antisocial personality. These were parental deprivation, socioeconomic status of the home of the rearing, broken homes, and personality disorders among the adoptive parents. It is unfortunate that these data were not available on every subject, because there is no way of knowing how representative the data are of the overall proband and control groups. However, they were available on a sizable subsample. These data did not demonstrate any significant differences, or even trends toward differences, between the 2 groups. Therefore, as best can be determined, the control group provided an adequate control for the probands on the above environmental variables.

An important point which must be considered is the question of whether any of the adoptive parents knew of the criminal background of the natural mother and the effect this knowledge might have had on the subsequent development of the adopted child. This is difficult to determine in a consistent manner in a study which spanned 30 years and included multiple adoption agencies. However, some information is available on those probands who were diagnosed as antisocial personality. Psychiatric and social histories were obtained on all of these probands from the various institutions they had contact with, and 5 of the 6 included interviews with parents. At no point in the records did this issue arise. Therefore, if the parents knew, it did not appear to figure prominently in the child's problem as the parent saw it.

When the final results become available, those probands diagnosed as antisocial personalities will be looked at closely in terms of their natural parents, their adoptive parents, early adoptive environment, preadoption experiences, and any other variables which may have been important in their outcome. Hopefully, this will help to answer some of the questions which remain at this point and perhaps provide some understanding of the relative role of genetics and environment in the development of antisocial personality.

References

Christiansen, K. O. Crime in a Danish twin population. *Acta Geneticae Medicae et Gemellologiae*, 1970, **19**, 323–326.

Cloninger, C. R., & Guze, S. B. Psychiatric illness and female criminality; the role of sociopathy and hysteria in the antisocial woman. *American Journal of Psychiatry*, 1970, **127**, 303–311.

Crowe, R. The adopted offspring of women criminal offenders: a study of their arrest records. *Archives of General Psychiatry*, 1972, **27**, 600–603.

Hollingshead, A. B., & Redlich, F. C. *Social class and mental illness: a community study*. New York: John Wiley & Sons, 1958, pp. 390–391.

Robins, L. N. *Deviant children grown up: a sociological and psychiatric study of sociopathic personality*. Baltimore: Williams & Wilkins, 1966.

Roe, A. The adult adjustment of children of alcoholic parents raised in foster homes. *Quarterly Journal of Studies on Alcohol*, 1944–45, **5**, 378–393.

Rosenthal, D. *Genetic theory and abnormal behavior*. New York: McGraw–Hill, 1970.

Schulsinger, F. Psychopathy, heredity and environment. *International Journal of Mental Health*, 1972, **1**, 190–206.

Skeels, H. M. Mental development in children in foster homes. *Journal of Consulting Psychology*, 1938, **2**, 33–43.

Yoshimasu, S. Criminal life curves of monozygotic twin-pairs. *Acta Criminologiae et Medicinae Legalis Japonica,* 1965, **31**, 5–6. Reviewed in E. Slater & V. Cowie: *The genetics of mental disorders,* London: Oxford University Press, 1971, p. 116.

7 REGISTERED CRIMINALITY IN THE ADOPTIVE AND BIOLOGICAL PARENTS OF REGISTERED MALE CRIMINAL ADOPTEES

Barry Hutchings and Sarnoff A. Mednick

Introduction

The recent renewed interest in biological etiologies for criminal behavior (e.g., Hare 1968) has stimulated a reevaluation of the genetic basis for criminality, for which the existing evidence is based almost wholly on twin studies, c.f., a recent summary in Slater and Cowie (1971). The present study is a preliminary report of a retrospective investigation using archival material of registered criminality among adoptees and their adoptive and biological relatives. Adoption is regarded as an *ex post facto* experiment in which the hereditary influences represented by the biological parents and the environmental influences as indicated by the adoptive parents can be separated. If there is a genetic basis for criminality then there should be an attendant correlation between the criminality of the biological parents and that of the adoptees. Furthermore, this correlation should be independent of the criminality of the adoptive parents.

The opportunity was available in Copenhagen of working with an adoptee material of unusually fine sampling characteristics. Kety, Rosenthal, Wender, and Schulsinger, with the support of contracts between Copenhagen's Psykologisk Institut and USPHS, established a file of 5,483 adoptees, encompassing all nonfamilial adoptions in the City and County of Copenhagen between 1924 and 1947. The creation of this file is de-

Barry Hutchings, M.Phil., Psykologisk Institut, Kommunehospitalet, Copenhagen.
Sarnoff A. Mednick, Ph.D., Psykologisk Institut, and New School for Social Research, New York.
This study was supported by USPHS grant M.H. 19225, The paper is based on a thesis by the first-named author accepted by the University of London for the Degree of M.Phil. 1972. The Adoptee Files used in this study were compiled by Kety, Rosenthal, Wender, and Schulsinger for their studies of schizophrenia. The design of the present study follows that of their work (1968). We wish to gratefully acknowledge permission to use these adoptee files.

scribed more fully in Kety, Rosenthal, Wender, and Schulsinger (1968). The study reported in the present paper centered on the 1,145 male adoptees in this file who were born between January 1, 1927 and December 31, 1941. Thus, the adoptees were between thirty and forty-four years of age at the time their criminality was ascertained in 1971. This would mean that the sample has passed through the greater part of the risk period for criminality, especially for registration of a first offense. It also has the advantage for epidemiological purposes of being a complete series, at least in the sense that it is unselected for criminality. Denmark provides excellent demographic facilities for such a study, with a small population having low rates of emigration and immigration, a national population register, and centralized police and psychiatric registers.

Initial Study of 1,145 Male Adoptees

For purposes of comparison the 1,145 adoptees were matched individually with a group of nonadoptees for sex, age, occupational status of their fathers, and residence (using old census lists).

The police records for Denmark are housed centrally in the Police Record Office (*Rigsregistraturen*). A chronological index by birth-date gives access to the main alphabetically arranged files on all persons "known to the police" (*Hovedkartotek*). In itself this file is unsuitable for criminological research as it includes very minor offenses and some things which are not offenses at all. Traffic offenses are included, although parking infringements, cycling without lights, and jaywalking are not. On the other hand certain administrative matters are registered, such as hospitalization with concussion and self-discharge of psychiatric patients from the hospital. Inclusion in this file cannot by itself be taken as an indication of antisocial behavior.

A separate criminal record (*Personalia Blad*) is kept on all persons who have at any time been convicted of offenses treated as *statsadvokatsager*. These correspond very closely to indictable offenses in British justice and can be contrasted with *politisager* (summary offenses). Generally speaking, indictable offenses are those against the Danish Penal Code (*Straffeloven*) plus a few of the Special Laws (*Saerlove*) such as narcotics offenses, cruelty to animals, and serious customs and excise offenses. On the other hand some offenses against the penal code, such as begging and disorderly conduct are normally dealt with as summary offenses. The distinction corresponds very roughly to the difference between felonies and misdemeanors in the United States.

It is the "criminal record" which is used as the primary source of information for this study. The lifetime risk for registration of a male in Denmark for criminality so defined is about 9 percent.

The 1,145 adoptees and their matched nonadopted controls were checked for registration in the police files. The distribution of these two

Table 1. Distribution of the Initial Sample of Adoptees and Their Matched
Nonadopted Controls in the Police Files

	Adoptees	Nonadoptees
Unidentifiable		1
Not known to the police	566	721
Minor offenses only	394	322
With criminal record	185	101
Total	1145	1145

groups in the files is shown in Table 1. It should be remembered that in this and succeeding tables the category "minor offenders" also includes some nonoffenders who are known to the police for one reason or another. As can be seen, 185 (16.2 percent) of the adoptees have criminal records, markedly more than either the controls or the corresponding population figure.

The adoptive and biological fathers of the adoptees and the fathers of the nonadopted controls were now checked through the police files. The distribution of the fathers is given in Table 2. The rates of criminality of the adoptive fathers and the fathers of the nonadopted controls are very similar. The biological fathers of the adoptees, however, evidence almost three times these rates of criminality.

The association between the registered criminality in the adoptees and their adoptive fathers is given in Table 3a. The percentage of adoptees with a criminal adoptive father increases from 9.2 percent of the unregistered adoptees to 21.7 percent of the criminal adoptees. ($X^2 = 19.52$ with d.f. = 2, p<.001.)

The criminal status of the 971 adoptees whose biological fathers could be identified is tabulated in Table 3b in relation to their biological fathers' criminality. Taking into account the increased criminality among the bio-

Table 2. Distribution in the Police Files of the Adoptive and Biological Fathers on the Initial Sample of Adoptees and of the Fathers of the Matched Nonadopted Controls

	Adoptive fathers	Biological fathers	Fathers of controls
Unknown or unidentifiable	26	174	27
Not known to the police	755	464	779
Minor offenses only	220	154	212
With criminal record	144	353	127
Total	1,145	1,145	1,145

Table 3a. Distribution of Adopted Danish Males Aged 30–44 Years by
Registered Offenses and by Criminality of Their Adoptive Fathers

Criminal record of adoptee	N	Criminal adoptive fathers (%)	Noncriminal adoptive fathers (%)
Not registered	554	9.2	90.8
Registered for minor offense only	385	14.0	86.0
Registered for criminal offense	180	21.7	78.3
Total	1119		

$x^2_2 = 19.52$, p < .001

Table 3b. Distribution of Adopted Danish Males Aged 30–44 Years by
Registered Offenses and by Criminality of Their Biological Fathers

Criminal record of adoptee	N	Criminal biological fathers (%)	Noncriminal biological fathers (%)
Not registered	473	31.1	68.9
Registered for minor offense only	334	37.7	62.3
Registered for criminal offense	164	48.8	51.2
Total	971		

$x^2_2 = 16.91$, p < .001

Table 3c. Distribution of Nonadopted Danish Males Aged 30–44 Years by
Registered Offenses and by Criminality of Their Fathers

Criminal record of nonadopted control	N	Criminal fathers (%)	Noncriminal fathers (%)
Not registered	706	9.5	90.5
Registered for minor offense only	314	12.4	87.6
Registered for criminal offense	100	21.0	79.0
Total	1,120		

$x^2_2 = 12.05$, p < .001

logical fathers themselves, the same pattern can be seen as with the adoptive fathers ($\chi^2 = 16.91$ with d.f. $= 2$, p<.001).

The results shown in Table 3c for the nonadopted controls are very similar to those of the adoptees and their adoptive fathers when expressed as percentages.

These results suggest an association between the criminality of the sons and their fathers. This association appears to approximately the same degree among both adoptees and nonadoptees. With the adoptees it appears on both the biological and adoptive fathers' sides.

Proband Study of 143 Criminal Adoptees and Matched Noncriminal Controls

The proband study was made in order to make a closer examination of the relatives of a criminal group and to compare them with the relatives of a noncriminal group. The design is discussed in Rosenthal (1970) and has been used by Kety et al. (1968) in their adoption study of schizophrenia. Schulsinger (1972) reported results using the same material as the present paper but considering the clinical diagnosis of psychopathy.

A group was selected from the 185 criminal adoptees to include all those cases where the biological fathers were identifiable and whose adoptive and biological fathers were born from 1890 onward. This latter criterion was to maximize the reliability of the police records. These 143 criminal adoptees became the index probands. They were matched individually with 143 adoptees who were not known to the police and who became the control probands. Matching was made on the basis of age and occupational status of the adoptive father. Table 4 presents identifying information on these two groups. As can be seen from the ages of first transfer to the adoptive homes, the amount of possible contact between the adoptee and his biological father was minimal. Actually in almost all cases it was nonexistent, since the transfer to the adoptive home took place from a children's home rather than from the biological home.

Thirty-three (23 percent) of the adoptive fathers of the criminal probands had received criminal records as against only 14 (9.8 percent) of the control adoptive fathers. ($\chi^2 = 8.25$ with d.f. $= 1$, p<.01). This difference was also reflected in various indices of criminality such as number of recorded cases and total length of sentence.

Among the biological fathers of the index cases there are 70 (49 percent) who have criminal records as against 40 (28 percent) of the biological fathers of the noncriminals ($\chi^2 = 12.42$ with d.f. $= 1$ p<.001). Again the difference was found with all measures of criminality considered. Not only were the index biological fathers more often criminal than the controls, but they were worse criminals, with an average of 7.1 cases recorded against each of the criminals among the index fathers compared with

Table 4. Mean Characteristics of Index Adoptees (Criminal Probands) and Their
Matched Control Adoptees (Noncriminal Probands)

	Index adoptees (criminal probands) (n=143)	Control adoptees (noncriminal probands) (n=143)
Occupational status of adoptive father	2.3	2.4
Occupational status of biological father	1.7	2.1
Age on 1 January 1971	35.3 yr.	35.3 yr.
Age at birth of child of		
Adoptive mother	30.8 yr.	31.6 yr.
Adoptive father	32.5 yr.	33.6 yr.
Biological mother	23.3 yr.	23.6 yr.
Biological father	26.1 yr.	27.9 yr.
Age at 1st transfer to adoptive home (median)	6 mth.	7 mth.
Age at legal adoption (median)	19 mth.	16 mth.
Income of adoptive father	DKr. 4290	DKr. 4387
Fortune of adoptive father	DKr. 2065	DKr. 1848

4.7 cases against the control fathers who were criminals. On both the adoptive and biological sides the differences were more marked with respect to property offenses than to crimes of violence, but due to the overwhelmingly greater incidence of property offenses and the difficulty of classifying individuals according to dominant offenses it is difficult to draw a hard conclusion here.

When the criminal adoptees are graded in terms of the severity of their criminality by classifying them as recidivists, prisoners, etc., there was some tendency for concordance rates to be positively associated with increasing severity. These differences were not statistically significant, but were more marked on the adoptive side.

While there are relatively fewer adoptive or biological mothers registered as criminals, their data are, proportionately in agreement with those of the fathers.

There was considerably more mental illness recorded for the index probands than for the controls with 56 of the former being known to the Psychiatric Register as against only 7 of the latter. Psychopathy was predominant, but it should be remembered that the presenting symptom is often the criminal act. More interesting in this context are the psychiatric records of the parents on the adoptive and biological sides. Here there was no difference between the psychiatric histories of the index and control adoptive parents. The adoptive parents had been quite thoroughly screened for mental illness and there was a notable absence of psychosis. Forty-seven of the index probands and 32 of the control probands had at least one *biological* parent who was recorded in the Psychiatric Register. This difference was statistically significant at a low level ($\chi^2 = 3.97$ with d.f. $= 1$ $p<.05$). The differences were mainly on the side of the mothers and were due to increased psychopathy and neurosis rather than psychosis. This data is given in Table 5.

Midwives reports were obtained on 92 of the index cases and 93 controls. The two groups did not differ in respect to the course of the pregnancy or delivery nor to the nature and extent of the ensuing complications. Slight differences that were noticed were in the direction of the controls having slightly more favorable obstetric histories.

Up to this point a number of variables thought to be of relevance in the determination of criminal behavior have been examined independently. Two questions stimulated the analyses reported in the following section: (1) What is the *combined* effect of the several predictors believed to be of relevance? and (2) Are different aspects of criminality differently determined?

The traditional way of attacking the first of these questions would be to combine several variables in "n-way" contingency tables. This is somewhat unsatisfactory. In the first place examination of more than two or three predictors in such a manner requires prohibitive numbers of cases, certainly more than are available in the present design. But basically the problem is that of spuriousness. To what extent does the effect of A on X disappear when the effect of B on X is taken into consideration? This of course can be extended to variables not being examined when it becomes a question of how much of the variance are we explaining? Such questions are best answered by the use of linear statistical techniques, in particular, multiple regression. These analyses enable one to measure the relation of the various predictors *inter se* and to avoid the risk of explaining the same variance several times and of giving the exaggerated impression of relationships which can arise from tabular analyses.

The independent variables were coded as contrasts, with a value of 1 signifying the presence of a particular factor, such as a criminal or psychiatric diagnosis, and 0 as its absence. The following variables are included to predict the criminality of the adoptee: Criminal record and psychiatric hospitalization in biological or adoptive mother or father, social class, severe birth complication, and several of the more promising interactions.

The data for the 143 noncriminal probands and 143 criminal probands were subjected to a stepwise multiple regression analysis.

The three factors which contributed significantly to the regression analysis were (and chosen in this order):

1. Criminality in at least one *biological* parent
   $F = 16.31$ with d.f. $= 1,284$ $p<.001$
2. Criminality in at least one *adoptive* parent
   $F = 8.04$ with d.f. $= 1,283$ $p<.01$
3. Psychiatric diagnosis for the biological mother
   $F = 5.88$ with d.f. $= 1,282$ $p<.05$

Table 5. Incidence of Psychiatric Illness in the Biological Parents of Criminal Index Probands and Noncriminal Control probands

	Index cases		Control cases	
	Biological mother	Biological father	Biological mother	Biological father
	22	32	10	23
With psychiatric history				
Primary diagnosis				
A1 Nongenetically predisposed psychoses		1		3
A2 Genetically predisposed psychoses				
a. psychoses of aging		1		1
b. manic-depressive psychoses		4		2
c. schizophrenia	1			
d. epilepsy				
A3 Psychogenic psychoses	1		4	1
B Neuroses	8			4
C Psychopathies	3	6	1	6
D+E Oligophrenia and other defects of ability		1	2	3
F Abnormal reactions	5	3		
Total persons given diagnoses	18	16	7	20
Alcoholism	1	5		8
Suicide (completed)	1	3		
Suicide (attempted)	3	5	2	5

It is realized that there are problems with stepwise procedures, and one cannot conclude from such an analysis that any one predictor is more important than another, or compare their relative importance. What one can do is to conclude that the predictors which make significant contributions to the regression equation do exert their effects independently of each other. In the present analysis the conclusion can be stated that the criminality of the biological parents and of the adoptive parents, together with the mental illness of the biological mother, make contributions to the criminality of the child even when the partial correlations of these variables among themselves have been taken into consideration. Taken together these three predictors reach a multiple R of 0.315 with the criminality of the adoptee indicating that altogether about 10 percent of the variance has been accounted for. When interpreting this figure one should recall that variables were coded as values of either 1 or 0 which would tend to underestimate the multiple R but also that the independent variable (criminality in the adoptee) represents two extreme groups from the population, namely, criminals and persons not known to the police, which would tend to overestimate the R.

When 12 indices of criminality among the criminal probands (such as type of crime and severity of sentence) were subjected to a factor analysis, 2 factors were extracted which together accounted for about 70 percent of the variance. Factor 1 which is much the larger, is loaded heavily on theft, property damage, multiple offenses, number of offenses and a relatively early start to the criminal career. It would seem to be a fairly general factor describing typical nonviolent criminality. Factor 2 is clearly a factor correlated with offenses against the person. The high loading on injuries to the victim contrasts with the low loading on thefts and property damage. The high positive loadings on ages at first and last offenses indicate that these are crimes occurring fairly late in life.

When the regression analysis and the factor analysis were subjected to a split-half replication they both turned out to be reasonably stable.

The two factors obtained in the factor analysis were each separately regressed on the predictors of the regression analysis in order to pose the question whether the different aspects of criminality were differently determined. This latter analysis was unclear in result and proved totally unstable when subjected to a split-half replication.

Cross–Foster Analysis

The cross-fostering method, well known in genetic studies with animals, has been described in the context of studies such as the present by Rosenthal (1970).

Returning to the total sample of 1,145 adoptees we find that the sample is just large enough to examine the cross-fostering situation. Fifty-two

Table 6. Cross-foster Analysis

	N	Status of adoptee		
		Not known to the police	Minor offender	Criminal
Adoptive father criminal but biological father not known to the police	52	26 (23.2)	20 (18.8)	6 (10.0)
Biological father criminal but adoptive father not known to the police	219	95 (97.8)	78 (79.2)	46 (42.0)
Total	271			

$x \frac{2}{2} = 2.50$ ns.

adoptees were born to biological fathers who were not known to the police but had criminal adoptive fathers. A larger group of 219 adoptees had criminal biological fathers but were adopted by fathers who were not known to the police. The distribution of these 271 adoptees in the police files is shown in Table 6. The expected frequencies in each cell are given in parentheses.

This can be taken as a direct test of whether having a criminal biological father is more important than having a criminal adoptive father, with respect to predicting criminality in the adoptee. It can be seen that the data is in the direction of the hereditary effect being more important than the environmental effect when indexed in this way, though the analysis is not statistically significant.

When neither the biological father nor the adoptive father is known to the police (n = 333) 35 or 10.4 percent of the adoptees are criminals. When both fathers are criminals (n = 58), then 21 or 36.2 percent of the adoptees are criminals. The percentages for the 2 cross-fostered groups above are respectively 11.2 percent and 21.0 percent.

Discussion

Within the limits of the adoption methodology there appears to be a correlation between criminality in adoptees and criminality in their biological parents. The most important limit of the adoption method is the possibility that the adoption procedure results in selective placement, promoting correspondence between the adoptive home and the characteristics of the biological parents. Indeed, in this study we noted a significant correspondence in the social classes of the biological and adoptive fathers ($r = 0.22$ with d.f. $= 880$; p<.001). The Danish organization which arranged many of the adoptions examined in this study states clearly that they do aim at matching in certain respects. Since the hallmark of the adoptive method is the separation of genetic and environmental influences this matching is a serious problem for the researcher.

From the incidence of criminality among adoptive fathers and among biological fathers one can calculate the expected number of cases where both fathers are criminal if only chance factors have operated in the matching at adoption. For the present sample of 1,145, allowing for the unidentifiable fathers, this expected figure is 55. The actual number of adoptees in the 1,145, both of whose fathers are criminal, is 58. Thus it can be concluded that whatever effect the matching of the adoption agency might have, it does not express itself directly in registration for crime of the fathers.

It is worth restating that the operational definition of criminal behavior used here is that of detected, apprehended and registered acts defined by the Danish society as worthy of police and judicial action. It is thus primarily an administrative definition. The extent to which it is also a sociological definition is debatable; the extent to which it is a biological definition is even more doubtful.

In considering whether or how these results, obtained in Copenhagen, Denmark, can be extrapolated to other national settings, one consideration should be borne in mind. The laboratory experimenter in behavior genetics reduces the variance ascribable to environmental influences when he wishes to explore the effects of strain differences. As environmental variance increases, the strain difference effects become more and more masked. While operating in a narrower range than is available to the laboratory researcher, the extent of variability of a natural, human research environment (in our case, Denmark) will also influence the strength with which genetic factors will become manifest. We would suggest that the amount of variability in a culturally and racially homogeneous population such as Denmark for almost any dimension will be less than that of a country like the United States. There is also the problem of extrapolation from an adopted sample to nonadoptees; criminality among adoptees may not be typical in certain respects. It follows, then, that in practical terms extrapolation of our Danish findings to other national situations must be conducted with great caution. Stated simply these findings should not be used to explain crime in the United States.

References

Hare, R. D. Psychopathy, autonomic functioning, and the orienting response. *Journal of Abnormal Psychology Monograph Supplement*, 1968, **73**(3, Pt. 2).

Kety, S. S., Rosenthal, D., Wender, P. H., & Schulsinger, F. The types and prevalence of mental illness in the biological and adoptive families of adopted schizophrenics. In D. Rosenthal & S. S. Kety (Eds.), *The transmission of schizophrenia*. Oxford: Pergamon, 1968, pp. 345-362.

Rosenthal, D. *Genetic theory and abnormal behavior*. New York: McGraw-Hill, 1970.

Schulsinger, F. Psychopathy, heredity, and environment. *International Journal of Mental Health*, 1972, **1**, 190-206.

Slater, E., & Cowie, V. *The genetics of mental disorders*. London: Oxford University Press, 1971.

8

DISCUSSION OF GENETIC STUDIES OF CRIMINALITY AND PSYCHOPATHY

Lee N. Robins

The opportunity to review these papers has contributed considerably to my education. For one who has followed the literature on the XYY chromosome and aggression only casually, the papers by Drs. Kessler and Razavi have greatly increased my recognition of the serious problems that confront efforts to decide the contribution of genetic defect to aggression. In the course of following the literature, I had recognized certain problems. For instance, I knew that it was unreasonable to compare karyotypes of babies with those of tall mentally defective prisoners, because one does not know whether the babies will be either tall or mentally defective. Therefore, the differences found in rates between babies and prisoners may have more to do with IQ and height than with aggressiveness. The papers presented today reveal other problems as well. First, there is the problem of mosaicism, so that laboratory results may be of doubtful validity. Second, changes in karyotype may occur over the lifetime and over historical periods, depending on the amount of radiation in the atmosphere and other such variables. As a result, newborns probably are not the appropriate group from which to get baselines. Comparisons should be based on age-matched, contemporary aggressive and nonaggressive samples. Third, I learned that abnormalities may be observable only in certain tissues, not necessarily in blood and skin, which are the most easily available ones. In addition to these difficulties, I learned one fact that might make the solution of the problem easier: there exists an association of cytogenetic abnormality with temporal lobe epilepsy and abnormal fingerprints. These defects can serve as indicators to allow choosing high-risk samples without having to go through the expense of karyotyping general populations.

It was interesting that Drs. Kessler and Razavi disagreed about what action they think appropriate on the basis of existing studies of cyto-

Lee N. Robins, Ph.D., Washington University School of Medicine, St. Louis, Missouri.

genetics and aggression. Dr. Kessler doubts their positive findings and wants to avoid alarming parents and thus creating self-fulfilling prophecies. Dr. Razavi, on the other hand, seems to recommend cytogenetic counseling. Whether to counsel or not should depend not only on the truth of the association between these mutations and aggression but also on what one has to offer the family with an XYY child predicted to be aggressive. Dr. Razavi believes not only the XYY and XXY chromosomes are associated with aggression, but he thinks he knows the reason and the proper treatment. He thinks the genetic abnormality causes problems of sexual identification and that these problems are then susceptible to psychotherapeutic management. This is a chain of argument for therapeutic intervention for which there seems little evidence in the data. First, although the rates of chromosomal abnormality seem to be high in his data on sexual offenders, many of the published studies report no real differences between sexual offenders and other criminals. Nor is there consistent evidence that offenders with abnormal genes outside of facilities for sexual offenders have any special sexual problems, or even that they are more aggressive than other prisoners. But even if we were to grant that they might be especially aggressive prisoners, we have no demonstrated successful method for treating them. The chief virtue of karyotyping them at present would seem to be to facilitate a possible legal plea of not being responsible for their aggressive acts. Even this argument is suspect, since we have no clear evidence for any special predisposition to aggression if the abnormal genes are not accompanied by temporal lobe epilepsy.

Dr. Kessler suggests that the association between institutionalization and aggression may be a spurious one, resulting from the association of both with low socioeconomic status. The fact that there are other correlates of both variables does not necessarily mean, however, that the relationship between them is spurious. For instance, XYY abnormalities may be one of the intervening variables through which low socioeconomic status causes aggression. That is, one of the effects of low socioeconomic status may be damage to the germ plasm which in turn causes the aggression. Dr. Kessler's chief argument against genetic counseling is that the vast majority of affected individuals are not in institutions for defective offenders. While I would take issue with his assumption that being out of an institution is tantamount to "successful adaptation to societal mores," I am also concerned that his very low risk figures result from confounding point prevalence data with lifetime prevalence. Even if only 0.4 percent of the XYY males in Scotland were hospitalized at any one moment in time, depending on their average stay, this may mean that the lifetime expectancy is twenty to thirty times that high, a risk great enough to warrant concern.

Yet I agree with Dr. Kessler that the evidence for the role of the XYY or XXY chromosome in aggression is still flimsy. However, the facts presented by these two papers suggest research techniques that might help answer this still unresolved problem.

1) Instead of newborns, one might use autopsy material both from the skin and from suspect sites in the brain to correlate these results with height, criminality, and IQ, using existing records from police and school for the latter two. Unlike newborns, the autopsy material will show evidence for XYY chromosomes, if it has ever existed, and one can also look for evidence in the areas of the brain in which it is hypothesized to affect behavior.

2) One could do cytogenetic studies of aggressive delinquents comparing those (a) from good and bad homes, and (b) those with defective fingerprints and those with normal fingerprints. If it requires a greater genetic predisposition to produce delinquents in a favorable home environment, one would expect the delinquents from good homes to be a high-risk group for genetic abnormalities. When one compares the frequency of XYY chromosomes in delinquents from good and bad homes or with and without normal fingerprints, one ought to also try to correlate any positive findings with their assaultive-nonassaultive histories.

3) One could use FBI fingerprint files to find out whether fingerprints of criminals are more often of the types associated with the XYY and XXY abnormalities than are fingerprints of noncriminals. The FBI files, of course, contain many of both types. As a novice in thinking about this kind of research, I would leave it to the authors to answer whether these suggested strategies seem reasonable. Meanwhile, my thanks to them for increasing my sophistication and thus, alas, my uneasiness with existing studies.

Finally, as an epidemiologist, I would like to remind them that even if one shows that abnormal genes are associated with antisocial behavior, one has not shown how important a contribution such genetic abnormality makes to the total pool of aggressive individuals. It may account for a very small proportion of those who concern us, as compared with the effect of ordinary inheritance or ordinary environmental factors.

Let me turn now to the papers by Crowe, and Hutchings and Mednick. These two papers are probably among the most exciting to have appeared in the field of psychopathy in many years. But these papers also have made me wiser and therefore sadder. The paper by Hutchings and Mednick has reinforced a basic principle that seems to appear over and over again in studies that attempt to separate nature from nurture. That principle is "Every ointment has its fly." The fly in the adoptive ointment is that this "natural experiment" is a pretty imperfect experiment. Hutchings and Mednick have shown a lack of random assignment by socioeconomic status to adoptive homes, something that we have all ignored in treating adoptive studies as though they completely separated genetic and environmental factors. With such a strong relation between socioeconomic statuses

of biologic and adoptive fathers, it is surprising that Hutchings and Mednick found no correlation between their criminality. In this country at
least, criminality and SES are so strongly correlated that one would
assume that a correlation with the one implied a correlation with the other.

Our own work (Robins 1966) with nonadoptees has shown that the
father's antisocial history was more important than that of the mothers.
This also seems to be true in Denmark, and for the same reason—criminality
in females is rare. The effect of criminality in the father is shown to be
greater than that of criminality in the mother for both biological and
adoptive parents. Crowe's work shows that the rare mother who is a felon
is highly influential, although since he does not know whether the biological father also has a criminal record, it is still possible that the real
influence comes from the father rather than from the mother. It may well
be that almost all felonious women are married to criminal men.

Since the mother is known to be a felon in Crowe's data and since other
studies suggest that therefore the father will also have a criminal record,
one would expect more criminality in the adoptive offspring studied in
Iowa than in those studied in Denmark. Gross rates show the opposite.
Recalculating from Table 5 in the Hutchings and Mednick paper, we find
that 28 percent of the sons of criminal biological fathers were criminal.
In the Iowa data, only 19 percent of the offspring of felon mothers were
criminal. However, figures are not really comparable because the Iowa
offspring were younger than the Danish offspring, and so may not yet
have become criminal, and they included both males and females. In addition we do not know whether record keeping in Iowa is as complete as it
is in Denmark. Perhaps the most important difference is that migration
out of Iowa is more probable than out of Denmark. It is quite possible
that many of the Iowa offspring have records in other states which were
not collected. It would be interesting to know what the rate of criminality
among the Iowa adoptees is when one restricts the sample to males over
twenty-five who are still living in that state. Each of these constraints
should tend to raise the rate in the Iowa sample and make it more comparable to the Danish sample, but perhaps the Iowa sample is not large
enough to allow doing this.

Dr. Crowe has ably underscored how arrests and psychiatric diagnosis
of psychopathy are intertwined. Since psychopaths seldom seek care,
those who come to psychiatrists have usually been sent as a result of their
contacts with the courts. Thus the population of people diagnosed psychopathic is in large part a subset of criminals, excluding the offspring
of well-to-do families who are trying to forestall judicial action by getting
a psychiatric diagnosis and treatment when legal difficulties threaten.

There is nothing in Crowe's paper to suggest that there exists anything
one would call a "psychopathic spectrum." Children of felons exceeded
controls only in the diagnosis of psychopathy itself or in the diagnosis

of "inadequate personality," which may be just a milder form of psychopathy.

Crowe's diagnoses were based largely on personal interview. In the Danish study, where diagnoses were obtained only from the psychiatric register, surprisingly, there was no elevation of psychopathy rates in the biologic fathers of criminal adoptees, although criminal probands received this diagnosis much more often than controls. One would assume that there must be the same association between criminality and psychopathy in the previous generation as in the current one. Criminal fathers' failure to appear in the psychiatric register suggests that only recently in Denmark have criminals been recommended to psychiatric attention.

In the light of the research by Cloninger and Guze (1970, 1973; Guze, Wolfram, McKinney, & Cantwell 1967) showing a high rate of hysteria in both female felons and the female relatives of male felons, it is surprising that neither study found an excess of hysteria in the females. The biological mothers of criminals in the Danish study do include four "neurotics," while there are none among mothers of control subjects. Are these hysterics? The Danish study also shows some increase in schizophrenia in both sons and biological fathers of criminals. One wonders whether this is the flip of the coin of Heston's (1966) finding of more psychopathy in the children of schizophrenics. It would be valuable to investigate the schizophrenic probands in the Danish study to learn whether they were the offspring of the schizophrenic parents, thus taking advantage of a major asset of this study, that one knows the psychiatric status of both biologic parents.

As Hutchings and Mednick have pointed out, the Danish study provides a marvelous collection of data usable for many purposes other than those to which they have so far addressed themselves. Let me mention one finding that struck me as particularly interesting. Biological fathers of adoptees are much more criminal than their sons (36 percent vs. 16 percent), while there is little difference in rates for fathers and sons among non-adoptees (11 percent vs. 9 percent). The biological fathers of adoptees are of course also more criminal than the adoptive fathers (12 percent). Further, biological fathers of *noncriminal* adoptees are more criminal than the biological fathers of criminal *non*adoptees (34 percent vs. 21 percent). One can conclude from these figures that criminality of the father is a more important determinant that his child will be adopted than it is of his child's behavior.

But why is there a falling off of criminality between generations among the adoptees? One plausible explanation is that the father's high rate has resulted both from his genetic and environmental factors, while for the sons, the latter have been removed by adoption. However, we must also remember a second way in which adoptive agencies interfere with the simple natural experiment: not only do they attempt to match adoptive and biological parents, but they screen out of the adoption process children with obvious physical and mental defects. If the biologic fathers of the adopted children also have nonadopted children in whom such defects are associated with later criminality, this could help to explain the falling off of criminality among the adoptees. It would be an elegant addition to a study like this one if nonadopted whole siblings of the adoptees could also be included.

The relationship between the criminality of the adoptive parents and their child is remarkably strong. While this suggests that environmental factors may be very important in the development of criminality, there are other possibilities. One of these is that the adoptive parents' arrests may be an effect rather than a cause of the arrest of the child. At least in this country, when a child is identified as delinquent, his family situation is reviewed and his parents sometimes cited for neglect. If an older offspring becomes involved in criminality, he may involve his parent directly or as an accessory to the crime because stolen goods are found in the house, etc. Thus it would be important not to assume that an association between crime in the adoptive parent and crime in the offspring always means that the adoptive parents' history is the cause of the child's.

These four papers are signs of a wonderful coming of age of the study of family effects on criminality. The time-worn sentimental phrase "there are no bad children, only bad parents" is finally being translated into a form in which it can survive scientific scrutiny. We have passed the point of blame and reached the point of examining mechanisms and circumstances under which the sins of the father are visited unto the seventh generation.

References

Cloninger, C. R., & Guze, S. B. Psychiatric illness and female criminality: the role of sociopathy and hysteria in the antisocial woman. *American Journal of Psychiatry*, 1970, **127**, 303–311.

Cloninger, C. R., & Guze, S. B. Psychiatric illness in the families of female criminals: a study of 288 first-degree relatives. *British Journal of Psychiatry*, 1973, **122**, 697–703.

Guze, S. B., Wolfgram, E. D., McKinney, J. K., & Cantwell, D. P. Psychiatric illness in the families of convicted criminals: a study of 519 first-degree relatives. *Diseases of the Nervous System*, 1967, **28**, 651–659.

Heston, L. L. Psychiatric disorders in foster home reared children of schizophrenic mothers. *British Journal of Psychiatry*, 1966, **112**, 819–825.

Robins, L. N. *Deviant children grown up: a sociological and psychiatric study of sociopathic personality*. Baltimore: Williams & Wilkins, 1966.

III

GENETIC STUDIES IN SCHIZOPHRENIA

9 SCHIZOPHRENIC SPECTRUM DISORDERS IN THE FAMILIES OF SCHIZOPHRENIC CHILDREN

Lauretta Bender

Previous reports have been made on 100 individuals who were recognized as schizophrenic in childhood at Bellevue Hospital from 1935 to 1952 and followed into the third to fifth decade when the diagnosis of schizophrenia was confirmed in all but 6 cases. It was shown that childhood schizophrenia was the early onset of a life course of schizophrenia of every possible type. Also, it was shown that a child might pass through phases of autism and functioning at a mentally retarded level; and typically prepuberty childhood psychosis, with remissions in puberty and early adolescence, followed by pseudopsychopathic and pseudoneurotic clinical states; and in adulthood the individual might show regressed autistic behavior and every form of adult schizophrenia. Or, as an adult, with the proper opportunity, the individual might make an adjustment in the community in spite of inadequate, dependent, withdrawn behavior with one or another mannerism or other schizoid trait; or the individual might be gifted and creative but erratic and queer (Bender 1970; Bender & Faretra 1972).

It was not difficult, then, when one saw or heard of any of this type of disorder in a relative to accept it as a possible form of schizophrenia. My study of these 100 individuals, studied in childhood and followed into adulthood, showed that they had a history of 215 mentally ill relatives—fathers, mothers, siblings, and collaterals—of whom 86 had a recorded diagnosis of psychosis in a hospital or medical agency and 129 had personality disorders sufficiently serious to require social or mental recognition (Bender 1970).

Meanwhile, the concept of schizophrenic spectrum disorders in the families of schizophrenics has been developed and summarized by Heston (1970) from the work of Kety, Rosenthal, Schulsinger, and Wender (Rosenthal & Kety 1968), including a reevaluation of the earlier work of Kallmann (1938, 1946), especially.

Lauretta Bender, M.D., Attending Psychiatrist, New York State Psychiatric Institute; Clinical Professor of Psychiatry, Ret., Columbia University.

In this study I have used 50 of the 100 cases of schizophrenic children followed-up into adulthood. I have excluded all cases in which the diagnosis of schizophrenia was not confirmed, and all cases born out of wedlock where no personal or family history of the putative father was available. There were several instances of 2 siblings of the same family; here I included only the oldest or first observed child and included the sibling in the spectrum. Otherwise the cases were taken in alphabetical order.

The 50 control cases were observed at Bellevue at the same time and under the same conditions as the children diagnosed schizophrenic. They had all been included in other clinical studies and also had been used as controls in the 1949–51 study of schizophrenic children, supported in part by NIMH (Bender, Freedman, Grugett, & Helme 1952). Then it was found that 25 percent of the controls were subsequently diagnosed schizophrenic at Bellevue, but were not so diagnosed by other hospitals or agencies at that time. Cases that were diagnosed schizophrenic by any one facility were not included among the controls. Cases were excluded where the birth was out of wedlock and there was no family history for the putative father. Siblings were included under one case. My follow-up on the controls was not as extensive as on the schizophrenics, where the follow-up extended to 1968.

My data on this report, like all my other studies, were not collected in a systematic research fashion although they were not intended for clinical purposes only. The staff on the children's service at Bellevue at that time was highly research-oriented, although this was for clinical studies. It included such persons as Paul Schilder, David Wechsler, Walter Bromberg, Abram Fabian, Alan Rapaport, Frances Cottington, and others. However, the material does not justify statistical validation. It has, however, been repeatedly reevaluated, presented at meetings, and published. I was personally involved in all the original and clinical research studies and follow-up studies.

The personal data of these two groups of subjects is shown in Table 1. There were 50 each of controls and schizophrenics. There was a similar sex distribution, 38 control males and 39 schizophrenic males indicating about one female to four males, a widely recognized distribution of the sexes in disturbed prepuberty children. They were observed at Bellevue at the same time, on the same ward, and under the same conditions. Their ages on first observation were similar: 4–10 years (4 years and ten months) to 13–3 years for the controls and 4–3 years to 12–9 years for the schizophrenics. At the last follow-up the controls were 18 to 27 years, much less than the schizophrenics who were 22 to 46 years in 1968. The IQ range for the controls was 60 to 120 with a mean IQ of 88, which is fairly typical for groups of problem children in any medical or social organization. The IQ's for the schizophrenics were typically more variable, ranging from "untestable" in many autistic children to the genius level of 153, and the mean IQ was 74.

Table 1. Schizophrenic Spectrum Disorder in Families of Schizophrenic
Children: Comparative Personal Data

	50 control subjects	50 schizophrenic subjects
Number of subjects	50	50
Number of males	38	39
Age when 1st observed at Bellevue, 1935 to 1952	4–10 to 13–3.	4–3 to 12–9 yrs.
Age (mean) 1st observed at Bellevue, 1935 to 1952	9–0	9–4 yrs.
Age at last follow-up	18 to 27	22 to 46 yrs.
Age (mean) at present time (if alive)	43	38 yrs.
IQ range as children	60 to 120	Untest. to 153
IQ mean as children	88	74
Number of subjects with IQ below 70	5	29
Number of subjects with adult adjustment in community	40	18
Ethnic religious background		
Jews	7	22
Catholics	31	15
Protestants	12	9
Mixed	0	4
Blacks	7	4
Social economic status		
Professional parents	1	11
Own business or adequate employment	8	6
Marginal or dependent	41	33
Diagnosis		
Schizophrenia, autistic onset, infancy	0	25
Schizophrenia, childhood onset, 2½ to 10 yrs.	0	25
Psychopath. behav. dis. (due to deprivation)	5	0
Primary behav. disorder, neurotic or conduct	17	0
Psychoneurosis, anxiety or obsess.-compul.	14	0
Organic brain disorder	14	0

Eighteen of the 50 schizophrenics, or a little better than one-third, were
in the community as adults in 1968, but for the most part in some sort of
dependent relationship and known as schizophrenics; a few were conspicuous for their accomplishments. Forty of the controls were in the community when last known, but the number is rather meaningless, since there
is no recent follow-up and several were in and out of courts or penal institutions. On the other hand, several were in the process of really normal
independent community adjustment.

The ethnic, religious, and social economic status was different for the
two groups. Among the schizophrenic children there were more parents
who were Jews and professional people. This has been much discussed
(Kanner 1949; Rimland 1964) to indicate a different family and life pattern
for childhood schizophrenia, especially autism. However, Szurek and his

coworkers did not find the same distribution of parents of children observed at the Langley Porter Clinic in San Francisco. Neither did we at Creedmoor State Hospital in the 1950's and 60's (Bender & Faretra 1972). It was our impression at Bellevue that the special distribution resulted from concerned and informed parents seeking out places for their deviately developing children in the 1930's and 40's when few such places were available.

There was a high percentage of children from low economic backgrounds, especially in the nonschizophrenic group. These were agency-referred during the depression years and were typical of the population on the children's ward of Bellevue at that time.

The different diagnosis in the 2 groups of children was, of course, by selection. Of the schizophrenic children, 25 had an onset of their schizophrenia with autism in infancy, and 25 had an onset from 2½ to 10 years of age after a relatively normal infancy or early childhood. They were all referred to Bellevue later, 4–3 years to 12–9 years.

Of the 50 control cases, none were ever diagnosed schizophrenia. Five were diagnosed psychopathic behavior disorders due to deprivation (institutional care in infancy and early childhood, a not uncommon condition in the 1930's and 40's). Seventeen were primary behavior disorders with neurotic traits or conduct disorders, usually from very disorganized homes. Fourteen had a psychoneurosis, usually of an anxiety or obsessive compulsive type. Fourteen had organic brain disorders, such as epilepsy, Sydenham's chorea, posttraumatic encephalopathy or an encephalitis.

The categories in the schizophrenic spectrum disorders were adapted as closely as possible to those offered by Heston (1970). They have been simplified to make the tables more simple and communicative, but mostly because of the difficulty in using these unsystematically collected data and in differentiating conditions between individuals seen long ago or known only through records.

1. Schizophrenia or dementia precox was used only when the individual received this diagnosis officially where he (or she) was under treatment.

2. Other diagnosis in hospital referred to individuals who were diagnosed as psychotics with epilepsy, mental deficiency, syphilis of the brain, alcoholism, or were criminals in a penal institution.

3. Mental retardation (IQ below 70) or organic brain disease, such as epilepsy, Sydenham's chorea, or Gilles de la Tourette syndrome when diagnosed officially by school clinic, social agency, outpatient service, or private physician but had never led to hospitalization.

4. Sociopathic personality, antisocial acting out in a disturbed individual recognized by a medical or social agency and destructive of the home life of the subject or the subject's parents.

5. Neurotic or inadequate personality, deviate personalities who were passive, dependent, nonproductive, immature, etc., and also recognized by medical or social agencies as destructive of normal home life.

In collecting the case studies for the controls, I had expected that some of them would have a family member or members with a diagnosis of schizophrenia, but I was in doubt about what I should do with such cases. Would they be considered members of a schizophrenic spectrum? Should I exclude such cases or treat them in some special category? However, there were no such instances. In the 1930's and 1940's at Bellevue on the children's service, we missed the diagnosis of schizophrenia in 25 percent of the cases that we selected as nonschizophrenic controls in 1950. When we did make the diagnosis we apparently included every condition now recognized as part of the spectrum. Thus we used such terms as pseudo-defective, pseudoneurotic, pseudopsychopathic as well as autism, schizophrenia of childhood, and remission states.

Parents (see Table 2). The schizophrenic subjects had 5 schizophrenic fathers and one diagnosed psychotic in an institution, 12 fathers with sociopathic personalities, and 19 with neurotic and inadequate personalities, making a total of 37 fathers of the possible 50 who were disturbed. The nonschizophrenic controls had 26 disturbed fathers, none of whom were schizophrenic, 4 otherwise diagnosed in hospitals or institutions, and 22 with deviate personalities. This is about a 2 to 3 ratio with the schizophrenic subjects, who had more schizophrenic and inadequate fathers.

The schizophrenic subjects had 6 schizophrenic mothers while the controls had none. Both the schizophrenic and control subjects had a relatively equal number of mothers who were institutionalized for other conditions than schizophrenia, were mentally retarded, or had organic brain disease without hospital care and were sociopathic personalities. But the control subjects had 15 neurotic or inadequate mothers while the schizophrenics had 7. In sum, the controls had 21 disturbed mothers and the schizophrenics had 23, which is not very different. The schizophrenic children had more schizophrenic mothers while the controls had more neurotic and inadequate mothers.

However, the total number of parents in the schizophrenic spectrum was 60 for the schizophrenic subjects and 47 for the controls, a ratio of 5 to 4. The difference is due to the larger number of schizophrenic and deviate personalities in the parents of schizophrenics. It will be noted that 60 percent of the parents of these schizophrenic children observed at Bellevue from 1935 to 1952 were disturbed, as were 47 percent of the parents of the nonschizophrenic children (also under study for other disturbed behavior).

There are two other bits of data concerning the parents that turn up in this study. When it is determined how many of the parents were out of the home during the childhood of the subjects because of death, institutional care, or desertion, it is found that 23 homes of the control cases were thus disrupted, as were the homes of 17 of the schizophrenic subjects, indicating that this was a big factor in both groups, but a bigger one for the nonschizophrenic control cases.

Table 2. Schizophrenic Spectrum Disorder in Families of Schizophrenic
Children: Data on Parents

	50 control subjects	50 schizophrenic subjects
Fathers		
Schizophrenia diagnosed in hospital	0	5
Other diagnosis in hospital or other institution	4	1
Nonhospitalized organic brain disorder or retardation	0	0
Sociopathic personality	10	12
Neurotic or inadequate personality	12	19
Total	26	37
Mothers		
Schizophrenia diagnosed in hospital	0	6
Other diagnosis in hospital or other institution	2	3
Nonhospitalized organic brain disorder or retardation	2	3
Sociopathic personality	2	4
Neurotic or inadequate personality	15	7
Total	21	23
Total parents	47	60
Parents out of home during childhood of subject due to death, institution, or desertion		
Fathers	20	10
Mothers	3	7
Total	23	17
Mothers, dominant and symbiotic, kept home intact throughout life of subject without help of inadequate, deserting, or psychotic father	2	17

A so-far unpublished study that I have made of the family patterns
of the original cases of the 100 childhood schizophrenics has shown that
the most common family pattern (30 percent) was that of an immigrant
family (mid-European Jews, Irish and Italian Catholics, and British West
Indian Blacks) in which the mother was dominant, adequate, and often
symbiotic with her children, especially with the schizophrenic one, and
kept the home intact and ready for her sick child to return to even as a
dependent adult, in spite of the lack of help from the father, who was
often inadequate, psychotic, or neurotic and was out of the home, often to
return to his dominant mother.

In this group of 50 schizophrenic subjects, 17 mothers were of this type.
There were 2 mothers in the control group who were somewhat similar.
Both had obsessive compulsive sons to whom they were overprotective,
one father was inadequate while the other was dead. Neither family were
immigrants.

Table 3. Schizophrenic Spectrum Disorder in Families of Schizophrenic
          Children: Data on Siblings

	50 control subjects	50 schizophrenic subjects
Siblings		
Schizophrenia diagnosed in hospital	0	12
Other diagnosis in hospital or other institution	14 (1 half)	9 (3 half)
Nonhospitalized organic brain disorder or retardation	3	0
Sociopathic personality	0	2
Neurotic or inadequate personality	3	4
Total disturbed siblings	20 (1 half)	27 (3 half)
Normal siblings (living through childhood of subject)	91 (6 half)	73 (10 half)
Total siblings	111 (7 half)	100 (13 half)

Siblings (see Table 3). The 50 schizophrenic subjects had 12 schizophrenic siblings, while the control subjects had none. The schizophrenic
subjects had 9 siblings (including 3 half-siblings) in institutions for other
conditions than schizophrenia, while the control subjects had 14 (including
one half-sibling) or about 50 percent more. The control subjects had 3
nonhospitalized siblings with mental retardation or organic brain disease,
while the schizophrenic subjects had none. The control subjects had 3
siblings with deviate personalities, while the schizophrenics had 6. The
schizophrenic subjects had a total of 27 disturbed siblings (including 3
half-siblings), while the control subjects had a total of 20, a ratio of about
4 to 3, the greater number for the schizophrenics being largely due to
schizophrenia in 12 siblings.

It should also be noted that the 50 schizophrenic subjects had 73 normal
siblings (including 10 half-siblings) living through childhood. Thus, including the 27 disturbed siblings, the schizophrenic children had 100 siblings (including 13 half-siblings). The control subjects had 91 normal
siblings (including 6 half-siblings) and 20 disturbed siblings with a total
of 111 siblings (7 half-siblings). The important difference is that the
schizophrenic subjects had more than 1 to 4 siblings who were disturbed
or belonged to the schizophrenic spectrum, while the control group had
less than 1 to 5 that were disturbed (though none were schizophrenic).

Collaterals (See Table 4). There is considerably more pathology recorded among the collateral relatives of the schizophrenics than the controls. Sixteen of the collateral relatives of the schizophrenics were known
to be schizophrenic, while the controls reported no schizophrenic relatives. Ten of the collateral relatives of the schizophrenics were in institutions for some other condition than schizophrenia, while this was reported

Table 4. Schizophrenic Spectrum Disorders in Families of Schizophrenic
          Children: Data on Collateral Relatives

	50 control subjects	50 schizophrenic subjects
Collateral relatives		
Schizophrenia diagnosed in hospital	0	16
Other diagnosis in hospital or other institution	6	10
Nonhospitalized organic brain disorder or retardation	0	3
Sociopathic personality	7	12
Neurotic or inadequate personality	0	14
Total	13	55

for 6 of the collateral relatives of the control group. None of the control
group had collateral relatives who had records of mental retardation or
organic brain disease outside of an institution, while 3 of the relatives
of the schizophrenic subjects did. There was a total of 26 collateral relatives of the schizophrenic subjects with deviate personalities and only 7
of the control group. Thus there were 55 disturbed collateral relatives of
the 50 schizophrenics and 13 for the control group, a ratio of 4 to 1. The
greater number of mentally disturbed relatives for the schizophrenic subjects was due to the greater number of disturbances in all categories of
diagnosis.

Discussion

This is a report of the schizophrenic disorders in families of schizophrenic children as compared to nonschizophrenic children as controls
who were under observation for other behavioral disorders at Bellevue
Hospital at the same time (See Table 5). To begin with, a family history
of mental illness was denied in 12 of the control families and in 3 of the
schizophrenic families for a ratio of 4 to 1.

There were 39 schizophrenic relatives (fathers, mothers, siblings, and
collaterals) of the 50 schizophrenic subjects and no schizophrenic relatives
of the controls. On the other hand, the controls had nearly the same number of relatives with another diagnosis than schizophrenia in a hospital
or other institution (26 and 23). Also, there were about the same number
of relatives for both groups with organic brain disorder and mental retardation who were not cared for in institutions. However, the schizophrenic
subjects had 30 relatives with sociopathic personalities and 44 with
neurotic and inadequate personalities, a total of 74, while the control
group had 19 relatives with sociopathic personalities and 30 with neurotic

Table 5. Schizophrenic Spectrum Disorder in Families of Schizophrenic
Children: Total Data on All Relatives by Diagnosis

	50 control subjects	50 schizophrenic subjects
Family history of mental illness denied	12	3
Schizophrenia diagnosed in hospital	0	39
Other diagnosis in hospital or other institution	26	23
Nonhospitalized organic brain disorder or retardation	5	6
Sociopathic personality	19	30
Neurotic or inadequate personality	30	44
Total	80	142

and inadequate personalities for a total of 49. There was a total of 142
mentally ill relatives for the schizophrenic subjects and 80 for the control
group, a ratio of 7 to 4.

The greater number of mentally ill relatives of the schizophrenic subjects is due to the actual cases of schizophrenia in all types of relationships and the greater number of sociopathic personalities and neurotic and
inadequate personalities. There is no significant difference between the 2
groups in the number of relatives who were in hospital or penal institutions with a mental illness other than schizophrenia, or with criminality,
or with organic brain disorder or mental retardation not cared for in institutions.

There is, however, a qualitative difference in the mental illnesses and
emotional disturbances of the relatives of schizophrenic subjects as compared to the control cases which is not well brought out in this study. It is,
of course, of the nature of schizophrenia itself. It is what makes it possible
to show figures in which there are a known number of schizophrenics
among the relatives in all categories of relationship to known schizophrenics, while none have appeared among the nonschizophrenic groups.

It should also be possible to define those characteristics of the sociopathic and neurotic and inadequate personalities that may be considered
schizoid, and thus perhaps different from the deviate personalities of the
control group; in other words, to differentiate those who belong to the
schizophrenic spectrum from those who do not and are relatives of the
control group. Something of this sort is suggested by the 17 dominant,
adequate, symbiotic mothers of schizophrenic children who could maintain
a home to which the schizophrenic child could return all his life in spite of
an absent, inadequate father. These 17 mothers were not included in the
schizophrenic spectrum and possibly should have been. There is also a
question of the gifted, creative, unusual personalities that are found

among the relatives of schizophrenics, i.e., the "superphrenics" of Karlsson (1968). Several occurred in this schizophrenic group, but none were mentioned among the relatives of the control group.

Conclusions

In the study of the family history of 50 schizophrenic children observed at Bellevue Hospital from 1935 to 1952, and of 50 control nonschizophrenic children observed for other problems at the same time and followed into adulthood, it has been shown that the schizophrenic children had 39 schizophrenic relatives in all categories of relationship, while the nonschizophrenic children had none. There were also 50 percent more sociopathic, neurotic, and inadequate personalities among the relatives of the schizophrenic subjects than among the control subjects. There was no difference in the number of relatives with other mental illnesses, mental retardation, organic brain disease, or criminality in or out of institutions. There was, however, a greater number of neurotic and inadequate personalities among the mothers of the nonschizophrenic children than among the schizophrenic (a ratio of 15 to 7, more than double).

Thus the schizophrenic spectrum, as evidenced by those subjects who were schizophrenic from childhood, would seem to comprise schizophrenia itself, and schizoid, sociopathic, and neurotic personalities.

References

Bender, L. The life course of schizophrenic children. *Biological Psychiatry*, 1970, **2**, 165–172.

Bender, L., & Faretra, G. The relationship between childhood schizophrenia and adult schizophrenia. In A. R. Kaplan (Ed.), *Genetic factors in schizophrenia*. Springfield, Ill.: Charles C Thomas, 1972, pp. 28–64.

Bender, L., Freedman, A., Grugett, A. E., & Helme, W. Schizophrenia in childhood: a confirmation of the diagnosis. *Transactions of the American Neurological Association*, 1952, **77**, 67–71.

Heston, L. L. The genetics of schizophrenia and schizoid disease. *Science*, 1970, **167**, 249–256.

Kallmann, F. J. *The genetics of schizophrenia*. New York: Augustin, 1938.

Kallmann, F. J. Genetic theory of schizophrenia. An analysis of 691 schizophrenic twin index families. *American Journal of Psychiatry*, 1946, **103**, 309–322.

Kanner, L. Problems of nosology and psychodynamics of early infantile autism. *American Journal of Orthopsychiatry*, 1949, **19**, 416–429.

Karlsson, J. Genealogic studies of schizophrenia. In: D. Rosenthal & S. Kety (Eds.), *The transmission of schizophrenia*. Oxford, England: Pergamon, 1968, pp. 85–94.

Rimland, B. *Infantile autism*. London: Metheum, 1964.

Rosenthal, D. & Kety, S. (Eds.), *The transmission of schizophrenia*. Oxford, England: Pergamon, 1968.

10

ON THE POSSIBLE MAGNITUDES OF SELECTIVE FORCES MAINTAINING SCHIZOPHRENIA IN THE POPULATION

Kenneth K. Kidd

Introduction

Two aspects of schizophrenia continue to intrigue geneticists: the inability so far to discern the mode of inheritance (Kidd & Cavalli-Sforza 1973) and the inability to identify factors maintaining the trait—apparently so evolutionarily detrimental—at such a high frequency (Huxley, Mayr, Osmond, & Hoffer 1964; Moran 1965). This latter problem is equally important for any mode of biological inheritance. Both the single-major-locus model and the multifactorial-polygene model for inheritance of schizophrenia appear to require selective mechanisms leading to a balanced polymorphism, models for which could be devised.

Falconer (1965) presented a threshold model for analyzing the genetics of multifactorial traits in which both genetic and environmental factors are obviously involved in the variation in the manifestation of the trait; the genetic component is generally considered to be polygenic. Gottesman and Shields (1967) applied it to data on schizophrenia with encouraging results. Models involving a single major locus for the genetic component of the variation in the population have also been developed for such traits (Elston & Campbell 1971; Cavalli-Sforza & Kidd 1972). From a purely statistical point of view, it is usually impossible to decide between the multifactorial (MF) and the single-major-locus (SML) models if the only data are average incidence rates among relatives of affected probands. For any trait classified as present or absent, such incidence rates always yield a heritability estimate for the liability on the MF model (Edwards 1969) and virtually always allow an acceptable range of parameter values on the SML model (James 1971). In neither case can the fit be tested. Recent work has con-

Kenneth K. Kidd, Ph.D., Department of Human Genetics, Yale University School of Medicine, New Haven, Connecticut.
I wish to thank Dr. T. Reich for his helpful discussions during this work.

firmed the equivalence of the two models in explaining the reported incidence figures for schizophrenia (Cavalli-Sforza & Kidd 1972; Kidd & Cavalli-Sforza 1973).

Other types of data are required if a distinction between the models is to be made. Reich, James, and Morris (1972) have shown that classification by severity makes discrimination at least theoretically possible, but adequate data of the required nature do not yet exist in the schizophrenia literature. Since active maintenance of schizophrenia in the population requires a selective advantage of heterozygotes (ignoring selective interaction with linked loci), such advantage may be a useful datum, if detectable, in discriminating between the multifactorial and SML models. A more thorough investigation was therefore undertaken. This paper presents an attempt to formulate the possible modes of action and magnitudes of selective forces on the SML model.

Because of the high heterogeneity in the values reported for the incidence of affected relatives of schizophrenics (Kidd & Cavalli-Sforza 1973), a caveat is necessary: no precise "correct" answer is possible. The analyses presented here use the "solutions" obtained by Kidd and Cavalli-Sforza (1973) as the basis. A different set of incidence values for the various types of relationships would lead to different numerical results, though, for reasons to be discussed, the qualitative conclusions would likely remain unchanged.

The Genetic Model

The general SML model postulates only two alleles, S_1, the normal allele, and S_2, the schizophrenogenic allele. There are three different, but interconvertible, formulations of the model. The one used by Elston and Campbell (1971) is based on the allele frequency of S_2 and a penetrance for each of the three genotypes. The formulation used by Morton, Yee, and Lew (1971) differs only in that it expresses the three "penetrances" as linear functions of three other parameters. The formulation of Cavalli-Sforza and Kidd (1972) is analogous to the multifactorial threshold model and is also based on four parameters: the allele frequency, q; the relative position of the heterozygote on the liability scale, h'; the environmental variance around each genotypic mean, ϵ^2; and the position of the threshold that divided the population into affected and unaffected, T. The scale for these last three parameters is defined by arbitrarily fixing the genotypic values of the two homozygotes on the liability scale; the present work sets S_1S_1 at 0 and S_2S_2 at 2. This SML model does require certain assumptions: random mating resulting in Hardy–Weinberg ratios of the genotypes, and absence of any correlation in environmental (nongenetic) factors among relatives. The model is otherwise very general and encompasses all other two-allele single-locus models from the Mendelian recessive to Mendelian dominant, including all intermediate formulations such as that of Slater (see Slater & Cowie 1971). In addition it allows for nongenetic factors to produce the phenotype in the absence of the schizophrenic allele.

On the MF model, two parameters, general incidence and incidence for any class of relationship, yield an estimate of the heritability of liability. On the SML model the general incidence and incidences for two classes of relationship (parent-offspring and sibling) combine to yield an infinite number of solutions (James 1971). All of these solutions are exactly equivalent in their predictions for the average incidences among relatives of the affected. However, an infinity of solutions is not necessarily hopeless. The added restriction that the allele frequency and all three penetrances must lie in the interval from 0 to 1 (required by the model, but not by the algebra) limits the range of acceptable solutions. Thus acceptable solutions (parameter sets) may only occupy a small part of the parameter space, though because the equations are continuous there would still be an infinite number of such parameter sets so long as there is any acceptable solution.

The analysis by James (1971) is based on equations by Kempthorn (1957) which relate general incidence, GI, and the additive, V_A, and dominance, V_D, genetic variances to the allele frequency, q, and the penetrances of the three genotypes (f_0, f_1, and f_2 for the genotypes with 0, 1, and 2 schizophrenogenic alleles, respectively). The variances, however, are not variances related to liability; they are based on the observed correlations calculated by arbitrarily coding the individuals 0 or 1 for absence or presence of the trait.

Assuming a value for q and solving Kempthorne's (1957) equations for the three penetrances, taking the positive roots, we obtain

$$f_2 = GI + \frac{p}{q} V_D^{1/2} + \left(\frac{2pV_A}{q} \right)^{1/2} \tag{1}$$

$$f_1 = \frac{GI}{p} - \frac{q}{p} f_2 + \left(\frac{V_A}{2pq} \right)^{1/2} \tag{2}$$

$$f_0 = (GI - 2pqf_1 - q^2 f_2)/p^2 \tag{3}$$

where $p = 1 - q$. If all three penetrances and the gene frequency lie between 0 and 1, the three parameters for the liability formulation can be calculated by first finding the three normal deviates, X_0, X_1, and X_2, above which the areas equal the three penetrances, f_0, f_1, f_2. The parameters T, h', and ϵ are then calculated on an arbitrary liability scale with the mean of the $S_1 S_1$ homozygote at 0 and the mean of the $S_2 S_2$ homozygote at 2:

$$\epsilon = \frac{2}{X_0 - X_2} \tag{4}$$

$$T = \epsilon X_0 \tag{5}$$

$$h' = T - \epsilon X_1 \tag{6}$$

Solutions for these six equations are graphed in Figure 1 as a function of q. It is obvious that quite a range of parameter values is possible; all sets of values, corresponding to each given q, predict the same incidences of affected relatives for *all* types of relationships, specifically, the values given in Table 1. There is thus no way to discriminate among these solutions with that type of data.

The model does embody heterogeneity that may be illuminating. Each genotype has both affected and unaffected individuals. Conversely, the

Fig. 1. The Equivalent Parameter Sets on the SML Model. For any given value of q the values of f_0, f_1, and f_2 are determined by equations 1–3. In the range of $q = .056$ to $q = .138$ (indicated by arrows on the ordinate) these can be interpreted as penetrances and T, h', and ϵ are defined by equations 4–6. The values in Table 1 are given by all parameter sets in this range. Note the different scales for the different curves: f_2, ϵ, and h' on the left; T, f_0, and f_1 on the right.

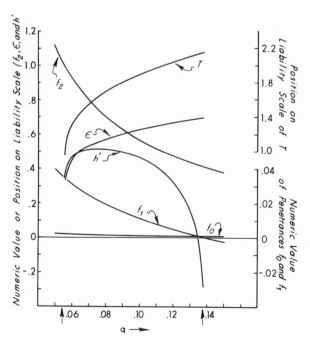

Table 1. The Incidences Among Relatives Predicted by the Equivalent
Parameter Sets Shown in Figure 1*

		Predicted incidence value
General incidence		.0087
Relatives of affected		
	MZ twin	.375
	DZ twin	.122
	Parent Offspring }	.052
	Aunt-Uncle Niece-Nephew }	.030
Specific matings		
	Affected by affected (S × S)	.341
	Normal by affected (N × S)	.052

*These values are in reasonable agreement with the average incidence values calculated by Kidd and Cavalli-Sforza (1973) and represent one of the "solutions" (set D) found by them. Their other sets give results similar to those presented here in Figures 1–5.

population of affected individuals is composed of representatives of all three genotypes, as is the population of normal unaffected individuals. Figure 2 shows how the population of schizophrenics would be constituted from the three genotypes, over the range of equivalent solutions. The variation is again considerable, but note that the S_2S_2 homozygotes never constitute more than 90 percent of the affected, while the S_1S_1 homozygotes never constitute less than 9 percent of the affected.

Fig. 2. The Proportions of Individuals of Each of the Three Possible Genotypes. This is plotted on the same ordinate as Figure 1; the concept is only defined for this range of q.

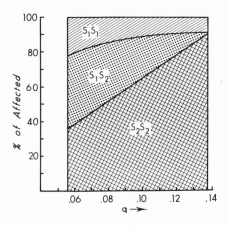

The Selection Model

In the past few years it has become increasingly evident that selection probably acts not separately on individual loci but on chromosome segments of many loci (Franklin & Lewontin 1970). The effects of such selection may not be predictable from fitnesses associated with only one of the loci involved. It may be that schizophrenia continues in the population because of a linkage disequilibrium with an entirely separate locus. However, at the moment only speculation along these lines is possible. A more precise quantification, within the limits of our genetic model, is possible by considering the single-locus selection model for schizophrenia.

At a locus at which relative fitnesses are independent of other loci, it is necessary that the average fitness of the heterozygote be greater than that of either homozygote. We have estimates for the decreased fitness of schizophrenics, but nothing allowing separate estimates for each genotypic group. Thus, we must assume all affected individuals have the fitness $1 - u$. (Were we able to separately estimate the fitness decrements for affected of each genotype, u would be the weighted average, using the relative proportion of each genotype (Figure 2) for the weights.) Using 1, $1 + h_1$, and $1 + h_2$ as the fitnesses of unaffected individuals of the three genotypes, the average fitnesses for the genotypes are $1 - uf_0$ for S_1S_1; $1 + h_1 - f_1(u + h_1)$ for S_1S_2; and $1 + h_2 - f_2(u + h_2)$ for S_2S_2.

At equilibrium

$$q = \frac{u(f_1 - f_0) - h_1(1 - f_1)}{u(2f_1 - f_0 - f_2) - 2h_1(1 - f_1) + h_2(1 - f_2)} \tag{7}$$

Solving for h_1 gives

$$h_1 = \frac{uq(2f_1 - f_0 - f_2) - u(f_1 - f_0) + qh_2(1 - f_2)}{(1 - f_1)(2q - 1)} \tag{8}$$

This can be solved for each parameter set (Figure 1) with the addition of values for u and h_2. Specific values are assumed for h_2. The value of u estimated by Slater and Cowie (1971), $u = 0.3$, is used, although Erlenmeyer–Kimling and Paradowski (1966) and Bodmer (1968) have shown that the selective disadvantage has decreased over the past few decades and was higher than 0.3 in the first half of this century. Figure 3 shows the solutions to this equation as a function of q (the penetrances being determined by q, see Figure 1) for the three different values of h_2. Again there is variation, though the results seem not too dependent on h_2 over most of this range. Most noticeably, the value of h_1 never exceeds 0.05, indicating that only a small average selective advantage for the unaffected heterozygote is required for a balanced polymorphism.

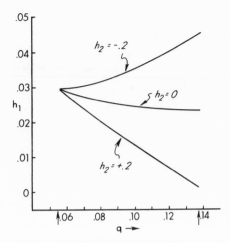

Fig. 3. The Selective Advantage h_1 of Unaffected Heterozygotes Required for a Balanced Polymorphism. These are plotted for three different values of h_2 as a function of q, which determines the other parameter values (see Fig. 1) used in equation 8. At the left, f_2 is 1 so h_2 is irrelevant; as a larger fraction of the S_2S_2 homozygotes are unaffected, h_2 becomes a more important parameter.

No method of detecting unaffected heterozygotes exists; so that only studies using unaffected relatives of schizophrenic probands are possible. Using the same methods, one can calculate the average proportion of unaffected relatives having each genotype. Weighting the fitness of each genotype by the frequency of that genotype among the unaffected siblings yields the curves in Figure 4. These represent the average fitness to be expected among unaffected siblings as a function of the gene frequency over the range of equivalent solutions. Note that the difference from 1 is always less than 0.04 and that in some circumstances the average fitness of unaffected siblings is actually less than 1.

Discussion

The results presented in Figures 1 to 3 show that the SML model yields considerable uncertainty with the type of data available. Nonetheless, the ranges of the variables and their interrelationships may be useful. It appears that the frequency of the schizophrenogenic allele is around 0.10 (approximate range $0.05 - 0.15$), certainly not a rare allele.

The uncertainty of the fit extends to the calculation of heritability, illustrating the problems encountered in using heritability in human genetics. Three entirely different "definitions" of heritability are possible for the same data. One is the multifactorial approach of Falconer (1965); a dichotomized bivariate normal distribution is assumed and the "heritability" of this liability is calculated. The values obtained are in the neigh-

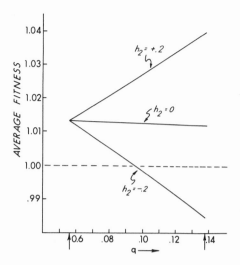

Fig. 4. Average Fitness of Unaffected Siblings of Schizophrenics. The three lines are the same as in Figure 3. Note that in some situations the average fitness falls below that for unaffected homozygous normals, even though the polymorphism is maintained by heterozygote advantage.

borhood of 80 percent (Cavalli-Sforza & Kidd 1972). A second approach assumes a single two-allele locus with affected individuals arbitrarily coded 1 and unaffected coded 0. Kempthorne (1957) gives formulae for calculating the additive, dominance, and total variance. These values (see equations 1–3) yield heritability estimates of 0.369 for the broad form and of 0.088 for the narrow. The third approach is to calculate the variances of the liability in the threshold formulation of the SML model. Kidd and Cavalli-Sforza (1973) give the formulae for calculating the broad heritability; the narrow heritability can be calculated from the following equations:

$$V_A = 2pq \left[2q + h' (p - q) \right]^2 \tag{9}$$

$$V_D = (2pq)^2(h' - 1)^2 \tag{10}$$

$$V_T = V_A + V_D + \epsilon^2 \tag{11}$$

$$h^2 = \frac{V_A}{V_T} \tag{12}$$

Both the broad and the narrow heritability are graphed in Figure 5. Note the great dependence of the results on the precise values of the parameters, even though all parameter sets give the same average incidence in relatives.

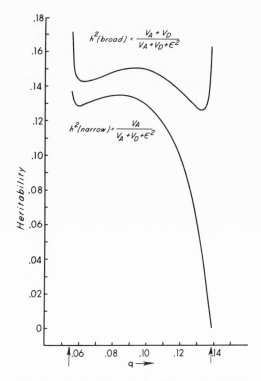

Fig. 5. Heritability of Liability on the SML Model. The variances are calculated according to formulae 9–12. The same ordinate is used here also, since all other parameters are determined by q and are only defined in this range.

It is obvious from this analysis that heritability is not an invariant value associated with a specific trait, but rather a function of the model assumed for the transmission of the trait. Lacking the ability to design selection and breeding programs to pragmatically test the values of heritability that are calculated, we must remember that in human genetics heritability estimates are always dependent on assumptions which are all too often unstated and which may be unwarranted.

The SML model as formulated in this analysis is able to quantitatively incorporate a nongenetic cause for a certain percentage of the affected. Such cases would be sporadic and nonfamilial and would certainly confound any attempt at general pedigree analysis. The specific results obtained do suggest that three "genetic" types of schizophrenics might exist; the familial histories of these three types should be different. Attempting by diagnosis to define categories of schizophrenia with different familial patterns is equivalent to using the two-threshold approach of Reich et al. (1972), if one assumes that severity is correlated with genotype. They showed that if any fit were possible to the SML model, it would be a unique solution with, of course, a margin of error. If more family studies are to be done, attempts will have to be made to classify individuals and relatives more clearly and to classify families as well. A biochemical marker would help in such a classification.

This SML selection model predicts a small increase in the fitness of the unaffected carriers of the schizophrenogenic allele. Depending on the fitness of homozygotes for that allele, the heterozygotes have an increased fitness at no more than 5 percent. Considering the unaffected siblings of schizophrenics, their average fitness, including all three genotypes, ranges from 0.98 to 1.04. There is thus the possibility of a decreased fitness even among the unaffected siblings. In such a case most of the selective advantage of heterozygotes would occur in families with no immediate relatives who are schizophrenic; these families could not even be identified. Thus, failure to demonstrate an increased fitness among normal siblings is not in itself a tenable argument against the SML model. Indeed, the maximum advantage expected on this model is sufficiently small that it would be difficult to demonstrate. However, there is the consideration that the values in Figures 3 and 4 are averages. If, as suggested by Heston (1970), many of the relatives are "schizoid," and have a reduced fitness, the increased fitness of the remainder would have to be greater. Also, the results of Erlenmeyer-Kimling and Paradowski (1966) suggest that there may be a large demonstrable advantage among unaffected siblings. In fact, the value they found, ~1.4, is far in excess of what this model would predict.

Conclusion

A further examination of the SML model for the inheritance of schizophrenia has shown that there is considerable uncertainty in the parameter values. There is, consequently, an uncertainty in the magnitude of selective factors that might be present. Some general conclusions are possible, however. If a SML is present, the frequency of the schizophrenogenic allele is about 10 percent (range 5–15 percent), and the average selective advantage of unaffected heterozygotes need be no more than 5 percent to maintain this allele in the population. Among the unaffected siblings of schizophrenics the maximum average selective advantage would be no more than 4 percent and it is possible that there would be a selective *dis*advantage of almost 2 percent. It would be difficult to demonstrate even a 4 percent selective advantage, and the possibility of no advantage exists.

Most of the ambiguity in estimation of the parameters is a direct result of the nature of the data analyzed: average incidence among relatives. Other types of data are needed to resolve the uncertainty. The SML model does, however, predict several types of heterogeneity that may be useful guides for further data collection. Affected individuals are collected from all three genotypes, suggesting that diagnostic criteria might be able to discriminate among them. Conversely, unaffected individuals of all genotypes occur, suggesting genetic heterogeneity among the unaffected relatives of schizophrenics. The possibility that some of these individuals show a slight reproductive disadvantage suggests that any reproductive advantage must be larger among the remainder. Thus, though present analyses are very ambiguous, many possible avenues for future research are suggested.

References

Bodmer, W. F. Demographic approaches to the measurement of differential selection in human populations. *Proceedings of the National Academy of Sciences*, Washington, D.C., 1968, **59**, 690–699.

Cavalli-Sforza, L. L., & Kidd, K. K. Genetic models for schizophrenia. *Neurosciences Research Program Bulletin*, 1972, **10**, 406–419.

Edwards, J. H. Familial predisposition in man. *British Medical Bulletin*, 1969, **25**, 58–64.

Elston, R., & Campbell, M. A. Schizophrenia: evidence for the major gene hypothesis. *Behavior Genetics*, 1971, **1**, 3–10.

Erlenmeyer-Kimling, L., & Paradowski, W. Selection and schizophrenia. *American Naturalist*, 1966, **100**, 651–665.

Falconer, D. S. The inheritance of liability to certain diseases, estimated from the incidence among relatives. *Annals of Human Genetics*, 1965, **29**, 51–71.

Franklin, I., & Lewontin, R. C. Is the gene the unit of selection? *Genetics*, 1970, **65**, 707–734.

Gottesman, I. I., & Shields, J. A polygenic theory of schizophrenia. *Proceedings of the National Academy of Sciences*, Washington, D.C., 1967, **58**, 199–205.

Huxley, J., Mayr, E., Osmond, H., & Hoffer, A. Schizophrenia as a genetic morphism. *Nature*, 1964, **204**, 220–221.

James, J. W. Frequency in relatives for an all-or-none trait. *Annals of Human Genetics*, 1971, **35**, 47–49.

Kempthorne, O. *An introduction to genetic statistics*. New York: John Wiley and Sons, 1957.

Kidd, K. K., & Cavalli-Sforza, L. L. An analysis of the genetics of schizophrenia. *Social Biology*, 1973, **20**, 254–265.

Moran, P. A. P. Class migration and the schizophrenic polymorphism. *Annals of Human Genetics*, 1965, **28**, 261–268.

Morton, N. E., Yee, S., & Lew, R. Complex segregation analysis. *American Journal of Human Genetics*, 1971, **23**, 602–611.

Reich, T., James, J. W., & Morris, C. A. The use of multiple thresholds in determining the mode of transmission of semicontinuous traits. *Annals of Human Genetics*, 1972, **36**, 163–184.

Slater, E., & Cowie, V. *The genetics of mental disorders*. London: Oxford University Press, 1971.

11 MENTAL ILLNESS IN THE BIOLOGICAL AND ADOPTIVE FAMILIES OF ADOPTED INDIVIDUALS WHO HAVE BECOME SCHIZOPHRENIC: A PRELIMINARY REPORT BASED ON PSYCHIATRIC INTERVIEWS

Seymour S. Kety, David Rosenthal, Paul H. Wender,
Fini Schulsinger, and Bjørn Jacobsen

The extent to which genetic factors operate in the transmission of schizophrenia has remained a controversial question. Evidence that the prevalence of schizophrenia is higher in the immediate families of schizophrenics than in the general population, and is especially high in their monozygotic twins, is not conclusive because of the difficulties in completely ruling out environmental influences and certain types of selective and subjective biases (Kety 1959; Rosenthal 1962). Family members and monozygotic twins may be expected to share many environmental influences in addition to their genetic endowment. Moreover, for a disorder such as schizophrenia, where the diagnosis cannot be established objectively and which shows a wide range in intensity and form, the occurrence of diagnosed schizophrenia in one member of a family is apt to enhance the discovery of mental illness in other members and, the likelihood of its being diagnosed as schizophrenia.

Seymour S. Kety, M.D., Psychiatric Research Laboratories, Massachusetts General Hospital, Boston.

David Rosenthal, Ph.D., and Paul H. Wender, M.D., Laboratory of Psychology, National Institute of Mental Health, Bethesda.

Fini Schulsinger, M.D., and Bjørn Jacobsen, M.D., Psychological Institute, Kommunehospitalet, Copenhagen.

This work was supported in part by grants from the National Institute of Mental Health, MH15602, the Schizophrenia Research Foundation of the Scottish Rite, Northern Jurisdiction, and the Foundations' Fund for Research in Psychiatry.

Studies with adopted individuals offer a means of minimizing these sources of error (Kety 1959). Since an adopted individual receives his genetic endowment from one family but his life experiences as a member of another, it may be possible to disentangle genetic and environmental factors in the development of schizophrenia, and studies by Heston and by us have utilized that device (Heston 1966; Kety, Rosenthal, Wender, & Schulsinger 1968; Rosenthal, Wender, Kety, Schulsinger, Welner, & Ostergaard 1968; Wender, Rosenthal & Kety 1968). If, in addition, a survey can be based on a total population of adopted individuals coming from natural families in which most of the schizophrenia occurs after the time of adoption rather than before, in which the schizophrenia was not a basis for offering the child for adoption, and where the mental status of the biological relatives and the adoptee are largely unknown to each other, it should be possible to reduce to a minimum the types of selective, ascertainment, and diagnostic bias alluded to previously.

Accordingly, in 1963 we began to collect a total sample of adults legally adopted at an early age by individuals not biologically related to them. Denmark was chosen because of its size, the relative homogeneity and stability of its population, and the excellent population and psychiatric records which exist there.* Two previous reports (Kety et al. 1968; Rosenthal et al. 1968) and the present study are concerned with the 5,483 individuals adopted by others than their biological relatives in the city and county of Copenhagen from the beginning of 1924 to the end of 1947. One of these (Kety et al. 1968) reported the prevalence of various types of mental illness in the biological and adoptive parents, siblings and half-siblings of 33 schizophrenic index cases who had been selected from that population of adoptees. For the purposes of the study, we had included as "schizophrenia" 3 subtypes defined in the *Diagnostic Manual of the American Psychiatric Association:* chronic schizophrenia, latent (ambulatory or borderline) schizophrenia, and acute schizophrenic reaction. The schizophrenic "index" adoptees were selected by independent review of abstracts of the institutional records of the 507 adoptees who had ever been admitted to a mental institution. These abstracts were prepared in English by a Danish psychiatrist and edited to remove any information regarding mental illness in other than the adoptee.

Unanimous agreement on a diagnosis of chronic schizophrenia, latent or borderline schizophrenia, or acute schizophrenic reaction was arrived at in 34 adoptees by 4 raters (F.S., D.R., P.H.W., S.S.K) on the basis of independent judgements followed by consensual agreement. As a control

*The authors are grateful to the State Department of Justice, to the officials responsible for the Folkeregister, and to the Institute for Psychiatric Demography (Drs. Annalise Dupont & Erik Strömgren), State Hospital, Risskov, for permission to use these files with appropriate safeguards regarding their confidentiality.

group, 34 probands* were selected from the adoptees who had not been admitted to a psychiatric facility by matching with each index case on the basis of age, sex, socioeconomic class of the rearing family, time spent with biological relatives, child-care institution, or foster home before transfer to the adopting family.

Our earlier report of prevalence and type of mental illness in the relatives was based simply upon an examination of the institutional records which were available for the 67 probands' biological and adoptive parents, siblings and half-siblings, whom we identified through the adoption records and the Folkeregister (a population register which permits one to trace a person's address, household, family, and children). We recorded independent diagnoses based on translated abstracts of these hospital records, edited to remove any biasing information which could suggest to us whether a subject was related to an index case or to a control, or was a biological or adoptive relative. We then conferred and arrived at a blind consensus diagnosis and broke the code. A significant concentration of what we called the "schizophrenia spectrum" of disorders was found among the biological relatives of index cases. In selecting the index cases we had been impressed with the gradations of severity in schizophrenia and had developed the hypothesis that there exists a broader range of schizophrenic disorder than is generally accepted. This spectrum included chronic, latent, and acute schizophrenia but also three less definite diagnoses (uncertain schizophrenia—chronic, latent or acute), and in addition, certain personality disorders which in the *A.P.A. Diagnostic Manual* are designated "inadequate personality" and "schizoid personality." The number of these illnesses which we found in the relatives was too small to permit a further breakdown of the schizophrenia spectrum into its components. Furthermore, we had secured practically no information about the environment of the probands other than the presence or absence of mental illness in the adopted relatives. One of our earlier studies (Rosenthal, et al. 1968) had also suggested that there were many more schizophrenics and individuals within the schizophrenia spectrum than those who had been hospitalized.

For these reasons, we felt that if we could carry out rather complete psychiatric interviews with these relatives, we would acquire considerably more information about their mental status and history, permitting a more exhaustive survey of the population with regard to schizophrenia and other psychiatric diagnoses and much more information about their life experience which might be correlated with the presence of schizo-

*The index cases were found to contain 2 who were monozygotic twins; because they were adopted by the same adoptive parents, these constituted 1 index case. A control proband had been chosen for each and both have been retained in the control sample. There are thus 33 index but 34 control families (biological and adoptive).

phrenia in the rearing family. We secured the collaboration of Dr. Bjørn Jacobsen, a Danish psychiatrist, who agreed to carry out the interviews and spent the greater part of the next two years in doing so. As a result of his conscientiousness and persuasiveness, he was able to secure voluntary interviews with 90 percent of the relatives who were alive and residing in Denmark or Scandinavia, since Dr. Jacobsen went to Norway and Sweden to interview the relatives who had emigrated there.

Table 1 indicates the present status of the relatives. By the end of February 1973, we had identified 512 parents, siblings and half-siblings of the 67 probands and they are distributed as shown: 173 biological index, 174 biological control, 74 adoptive index, and 91 adoptive control relatives. Among the relatives who had died, there is a very significantly higher proportion of the index biological relatives than their controls. We are in the process of analyzing the death certificates and medical records on all the relatives who have died to seek an explanation of this interesting difference. There is a higher incidence of suicide and accidental death among the index biological relatives, but whether this will account for all the difference remains to be seen.

Among the other unavailable relatives, those who emigrated beyond Scandinavia, disappeared, or were institutionalized, there is a random distribution. Only 10 percent of the available subjects would not agree to be interviewed, but these show no significant concentration in any one group. Even though a subject refused initially, Dr. Jacobsen would telephone and try to persuade him, and, occasionally, when he passed his door, would knock and attempt personally to obtain his cooperation, and with considerable success. In addition, in 12 instances, even though the individual persistently refused to give an interview, Dr. Jacobsen nevertheless obtained considerable information in the process. In 23 subjects, however, he obtained insufficient information to permit a satisfactory evaluation of mental status. Thus interviews were completed with 90 percent of the available subjects and sufficiently completed with an additional 4 percent, and those who did not participate are randomly distributed.

These interviews were extremely exhaustive, 35 pages in length, including many check lists and much narrative material, and covered the major aspects of the life experience: sociological, educational, marital, occupational, and peer relationship history from birth, medical background, and a careful mental status examination.

These interviews were transcribed in English (Dr. Jacobsen conducted them in Danish, but dictated them in English), and the transcripts were edited to remove any clues which a sophisticated reader might use to guess that this was a biological or adoptive relative of an index case or a control. The edited interviews, from which all information suggesting the relationship of the person interviewed to a proband or to mental illness in the family had been removed, were then independently read by Rosenthal,

Table 1. Status (as of February 1973) of the Biological and Adoptive Relatives
(Parents, Siblings, Half-siblings) of the Index and Control Probands

	Biological relatives		Adoptive relatives		Total relatives
	Index	Control	Index	Control	
Total identified	173	174	74	91	512
Died*	35	13	35	36	119
Left Denmark Sweden or Norway	13	11	0	2	26
Disappeared	1	1	0	1	3
Alive and accessible	124	149	39	52	364
Agreed to interview	112	138	34	45	329
Refused interview	12	11	5	7	35
Refused but adequate information	6	2	1	3	12
Refused and inadequate information	6	9	4	4	23
Interview or adequate information obtained	118	140	35	48	341

*The only significant differences between the groups are the number of deaths being
significantly higher for the index biological relatives vs. control (p=0.0004) and for control
adoptive us. control biological relatives (p<0.0001); the latter is undoubtedly a reflection
of the age differences. The percentage of interviews granted or refused in the accessible
populations are not significantly different.

Wender, and me. Each of us recorded a primary psychiatric diagnosis; later
a conference was held in which we were required to come to a consensus
on the diagnosis of each subject. After that diagnosis was recorded, the code
was broken and the subjects allocated to their respective groups.

With regard to psychiatric diagnoses outside the schizophrenia spectrum, there is a trend toward a lower proportion of these in the biological
relatives of schizophrenics than in the control relatives. This, however,
would come about if we found more schizophrenic illness in the biological
relatives and if those diagnoses were to override any other diagnosis we
might have made. In Table 2, therefore, we have removed those with
schizophrenia spectrum diagnosis and tabulated the prevalence of other
psychiatric diagnoses in the remainder. Approximately 37 percent of these
subjects were diagnosed as normal, or without psychiatric diagnosis.
Organic illness occurred in 6 percent of the biological and 13 percent of the

Table 2. Psychiatric Diagnoses Outside the Schizophrenia Spectrum Made by a Consensus of Three Raters in Edited Interviews

	Biological Relatives				Adoptive Relatives			
	Index	Prevalence (%[a])	Control	Prevalence (%[a])	Index	Prevalence (%[a])	Control	Prevalence (%[a])
Total identified	173		174		74		91	
Complete or adequate interviews	118		140		35		48	
Interviews without schizophrenia spectrum diagnosis	81		121		31		41	
Normal	30	(36.6)	49	(40.5)	11	(35.5)	11	(26.8)
Psychiatric diagnosis: Organic	7	(8.5)	6	(5.0)	5	(16.1)	6	(14.6)
Neurosis	4	(4.9)	6	(5.0)	3	(9.7)	2	(4.9)
Affective disorder	2	(2.4)	11[b]	(9.0)	1	(3.2)	3	(7.3)
Personality disorder	27	(32.9)	39	(32.2)	8	(25.8)	15	(36.6)
Psychiatric diagnosis other than schizophrenia spectrum	40	(48.8)	62	(51.2)	17	(54.8)	26	(63.4)

[a]Calculated as percent of interviewed relatives, excluding those with schizophrenia spectrum diagnosis.
[b]Prevalence of affective disorder is lower in index biological relatives than in the control biological relatives (p=0.049); for none of the other diagnostic categories is the prevalence significantly different between index relatives and their respective controls.

adoptive relative groups. This is largely cerebrovascular disease and the difference reflects the fact that the adoptive parents are older than the biological parents. Neuroses are randomly distributed and rather few. Affective disorder was diagnosed less frequently in the biological relatives of index cases than in the biological relatives of their controls, whereas personality disorders show no tendency to cluster in one group of relatives. The total of all psychiatric diagnoses presumably unrelated to schizophrenia is fairly evenly distributed, approximately 50 percent in both the biological index and the biological control relatives.

If we consider the total schizophrenia spectrum (Table 3), we find that 21 percent of the biological index relatives fell within that category, as compared to 11 percent of the biological relatives of the controls, and 5 and 8 percent respectively of the adoptive relatives. There is thus a highly significant concentration of schizophrenia spectrum disorders in the biological relatives of index cases. It is interesting that in the previous study, based only upon institutional records, we found a total of 21 people in the schizophrenia spectrum. Now we find 67, an increase by more than threefold, confirming the previous indication that there is considerably more schizophrenia-related illness in the population than reaches the attention of psychiatric institutions.

Now we have enough cases to permit us to break down the schizophrenia spectrum and test the relative prevalence and genetic relationship of its components. Table 3 also gives us these categories, the first of which is the definite diagnosis of schizophrenia, chronic or latent. No definite diagnosis of acute schizophrenic reaction was made in any of the relatives. That is not surprising in view of our fairly restrictive criteria for making that diagnosis—one or two acute schizophreniform episodes without clear evidence of premorbid or residual schizophrenic psychopathology. An interview at a later date is much more likely to find existing psychopathology than a brief episode in the past.

In our previous study based only on institutional records, we found 11 cases of definite schizophrenia in the total sample of relatives. Now we find 17, or 1.5 times as much, a ratio less than that for the total schizophrenia spectrum, which suggests that with severe mental illness one is not as likely to remain out of contact with psychiatric institutions.

There is a high and statistically significant concentration of definite schizophrenia in the biological relatives of the adopted index cases but not in their adoptive relatives. This is also true for our diagnosis of uncertain schizophrenia. Of course, the calculated prevalence varies according to the population denominator used. Should one use as a denominator the number of interviews that were conducted, the relatives who are alive, or the relatives in whom any diagnosis was made? We feel that the most unbiased and conservative denominator is simply the total number of identified relatives in each group, and that is what we have used. On that

Table 3. Prevalence of Schizophrenia Spectrum Disorders in the Biological and Adoptive Relatives of Schizophrenic Index and Control Probands (from Consensus Diagnosis on Interview)

Type of relatives	Number identified	Number interviewed	Total in schizophrenia spectrum		Schizophrenia spectrum						
					Schizophrenia			Uncertain (D)	Total B&D		Schizoid inadequate personality
					Definite (B)						
					Chronic B1	Latent B3	Total B1&B3	D1,D2,D3			
			N	%[a]	N (%)	N (%)	N (%)	N (%)	N (%)	N (%)	N (%)
Biological index	173	118	37	(21.4)	5 (2.9)	6 (3.5)	11 (6.4)	13 (7.5)	24 (13.9)	13 (7.5)
All biological controls	174	140	19	(10.9)	0 (0)	3 (1.7)	3 (1.7)	3 (1.7)	6 (3.4)	13 (7.5)
Screened biological controls[b]	113	86	11	(6.4)	0 (0)	1 (0.9)	1 (0.9)	0 (0)	1 (0.9)	10 (8.8)
p[c] (all index vs. all controls)	n.s.	n.s.	0.006		0.03	0.25	0.026	0.009	0.0004	n.s.
p (all index vs. screened controls)	n.s.	n.s.	0.007		0.08	0.16	0.019	0.001	0.00003	n.s.
Adoptive index	74	35	4	(5.4)	1 (1.4)	0 (0)	1 (1.4)	1 (1.4)	2 (2.7)	2 (2.7)
All adoptive controls	91	48	7	(7.7)	1 (1.1)	1 (1.1)	2 (2.2)	3 (3.3)	5 (5.5)	2 (2.2)
Screened adoptive controls[b]	64	34	3	(4.7)	0 (0)	0 (0)	0 (0)	1 (1.6)	1 (1.6)	2 (3.1)
p (all index vs. all controls)	n.s.	n.s.	n.s.		n.s.	n.s.	n.s.	n.s.	n.s.	n.s.
p (all index vs. screened controls)	n.s.	n.s.	n.s.		n.s.	n.s.	n.s.	n.s.	n.s.	n.s.

[a] In each instance percent is calculated as N/identified relatives.
[b] Relatives of the 23 "screened" controls (i.e., interviewed and found to be free of any suggestion of schizophrenic illness).
[c] p in this and other tables refers to Fischer's exact probability.

basis, schizophrenia, definite or uncertain, was found in 13.9 percent of the biological relatives of index probands compared with 3.4 percent in the biological control relatives and a similar low prevalence in both groups of adoptive relatives.

Table 3 also shows the distribution of consensus diagnoses of inadequate or schizoid personality in the identified relatives, and these do not show a concentration in the biological index relatives. Although our earlier studies (Kety, et al. 1968; Rosenthal, et al. 1968) did not develop evidence that those diagnoses per se significantly differentiated biological relatives of schizophrenics from controls (they were simply part of the spectrum which did), we are not prepared to dismiss the possibility that there is a schizoid or inadequate personality which is genetically related to schizophrenia. One of the three judges (D.R.) did make such a differentiation successfully ($p<.05$), and there are other data from our adoption studies in Denmark (Rosenthal, Wender, Kety, Schulsinger, Welner, & Jacobsen 1974) to suggest that schizoid personality in one parent, where the other parent is more clearly schizophrenic, enhances the likelihood or the intensity of schizophrenic illness in the progeny. In the present study, if we limit the examination to first-degree relatives or if we include only those inadequate or schizoid personalities where at least one rater made a stronger diagnosis, we find a significant concentration in the biological index relatives. This is simply to indicate the likelihood that within our consensus diagnosis of inadequate or schizoid personality there may be a type genetically related to schizophrenia which we cannot yet define precisely enough to separate from other nonspecific personality deviations.

Table 3 also includes the biological and adoptive relatives of "screened" controls who show a prevalence rate for these disorders which is considerably less than that for the control relatives as a whole. We had selected the 34 controls from the 4,976 adoptees who had no record of ever having been admitted to a psychiatric facility, but we were not sure that these were paragons of mental health. In the course of the interview study, and unbeknownst to Dr. Jacobsen, we included the 34 control probands among the subjects and Dr. Jacobsen interviewed all but 8 of them. The refusal rate among the available control adoptees was about what it was in the rest of the population (2/28). These were included in the interviews we evaluated and diagnosed without our differentiating them from the others. Nine of these we called normal, one was diagnosed neurotic, one as affective disorder, and 12 were called personality disorders other than schizoid or inadequate personality. These 23 controls we designate as "screened" to indicate that there was no suggestion of schizophrenic disorder among them. The 11 remaining controls included 3 who were schizoid or inadequate personalities, 2 who were dead, 4 who emigrated and 2 who refused to be interviewed: these we call "uncertain." There is a significantly lower prevalence of definite and uncertain schizophrenia among the biological relatives of these screened controls than among the biological relatives of the controls taken as a whole, and the prevalence of definite and uncertain schizophrenia in the biological index relatives is in marked contrast to the absence of these disorders in the biological relatives of the controls so screened (13.9 percent vs. 0.9 percent).

Table 4 lists the consensus diagnoses of schizophrenia (definite or uncertain) which were made in any relative either on the basis of the interviews or in our earlier study based on institutional records. Combining the two independent sources of information yields the maximum ascertainment of schizophrenic diagnosis in all groups and reinforces their concentration in the biological relatives of the index cases.

This evidence is compatible with a genetic transmission for schizophrenia, but it is not entirely conclusive, since there are possible environmental factors such as *in utero* influences, birth trauma, and early mothering experiences which have not been ruled out. One cannot, therefore, conclude that the high prevalence of schizophrenic illness found in these biological relatives of schizophrenics is genetic in origin. However, the largest group of relatives which we have is, understandably, the group of biological paternal half-siblings. Now, a biological paternal half-sibling of an index case has some interesting characteristics. He did not share the same uterus or the neonatal mothering experience, or an increased risk of birth trauma with the index case. The only thing they share is the same father and a certain amount of genetic overlap. Therefore, the distribution of schizophrenic illness in the biological paternal half-siblings is of great interest. Table 5 contains the relevant findings.

There are 63 biological paternal half-siblings of index cases and 64 paternal half-siblings of controls, which have between them 16 individuals diagnosed by records or interviews as definite or uncertain schizophrenia. There is, however, a highly unbalanced distribution; 14 are among the half-siblings of index cases and only 2 in the controls (p=0.001). If restricted only to definite schizophrenia, chronic or borderline, there are a total of 9 such cases, 8 in the index group, 1 in the control group (p=0.015). Among the 42 biological paternal half-siblings of the screened controls, there is no case of definite or uncertain schizophrenia. We regard this as compelling evidence that genetic factors operate significantly in the transmission of schizophrenia.

We have dealt up to this point with these relatives as individuals; but, as a matter of fact, they are members of families and it is probably not entirely appropriate to perform statistical analyses using the degrees of freedom represented by the individual relatives when these are clustered into a limited number of biological and adoptive families. Therefore, we have also analyzed these data in terms of families.

Table 4a and 4b give the distribution of definite and uncertain schizophrenia in the biological and the adoptive families of the index cases and the controls for those probands who have at least one such diagnosis among their biological or adoptive relatives, and Table 6 summarizes the information by type of family. Whereas the number of adoptive families with at least one such case is relatively small and not significantly different for index vs. control probands, more than half of the biological families of the index cases (17/33) show one or more definite or uncertain schizophrenias as compared with 5 of 34 biological control families and only one biological family of the 23 "screened" controls. There are significant differences in this regard between the index families and the combined control families (p=.0014), or the "screened" control families (p=.00013)

Inspection of Table 4a reveals some clusterings which are interesting from the genetic point of view. Chronic schizophrenia was found only in the biological relatives of index probands with chronic schizophrenia. There is an unexpectedly high prevalence of schizophrenia in biological half-siblings of index probands when the shared parent was also schizophrenic. These and many other questions will be examined in greater detail in a later, more complete report.

Some questions arise in the evaluation of these data. The first involves inter-rater reliability. A complete analysis of our individual diagnoses and the degree of agreement among them will be presented elsewhere. In formulating the consensus diagnoses we were constantly impressed with the agreement among our independent diagnoses. If we had depended on any single rater, the results would have been essentially the same. A statistically significant concentration of definite and uncertain schizophrenia was found by each rater, but only in the biological relatives of the index cases. We differed, as was indicated earlier, in the frequency with which we made the diagnosis of schizoid or inadequate personality and in our ability to differentiate the index from the control biological relatives on that criterion.

A more serious question concerns the objectivity of the diagnoses or their freedom from bias on the basis of a preconceived genetic or environmental point of view. It was relatively easy for the raters to remain blind with the help of extensive editing of the interview transcripts by an independent person who tended to err more in deleting than including information which might bias us. But one problem that we could not avoid is the possible bias that Dr. Jacobsen may have developed in his intensive interviews with subjects who may have volunteered information about adoption or mental illness in a relative, even though he did not inquire about these matters. Although we could not avoid this possibility, we did obtain information about it when it occurred. Dr. Jacobsen was asked to, and conscientiously did, note in a designated part of the inter-

Table 4a. Consensus Diagnoses Based upon Institutional Records or Interviews of Schizophrenia (Definite: Chronic B1, Acute B2, Latent B3, or Uncertain: Chronic D1, Acute D2, Latent D3) in the Biological and Adoptive Families of Schizophrenic Probands

Proband	Diagnosis	Biological family					Adoptive family			
		Mother	Maternal half-sibs	Father	Paternal half-sibs	Full sibs	Mother	Father	Half-sibs	Full sibs
S3	B1	$1\frac{\overline{B3}}{\overline{B1}}$	$1\overline{B1}$	1	$3^{a}\frac{\overline{D1}}{d^{c}}$	0	1	1	0	0
S5	B1	1	1	1	2	$1\overline{B3}$	1	1	1	1
S6	B1	1	2	$1\frac{\overline{D3}}{r}$	4	1	1	1	0	1
S8	B1	1	0	1	$2\overline{D3}$	0	1	1	0	0
S9	B1	1	0	1	$1\overline{B3}$	0	1	1	0	0
S18	B1	1	6	$1\overline{B3}$	$6\frac{\overline{B3}\ \overline{B3}\ \overline{B3}}{\overline{B1}\ \overline{B1}\ \overline{D3}}$	0	1	1	0	0
S22	B1	$1\overline{B3}$	$2\frac{B1}{\overline{B1}\ \overline{D3}}$	3	12	0	1	1	0	0
S25	B1	$1\overline{D3}$	$1\frac{\overline{D3}}{e}$	1	$1\overline{B3}$	0	1	1	0	$1\overline{D3}$
S31	B1	1	$2\overline{D3}$	1	0	0	1	1	0	0
S32	B1	1	0	1	$2\ \overline{D3}\ \overline{D2}$	0	1	$1\frac{\overline{D3}}{d}$	1	0
Remainder[b]	7B1	7	11	7	7	0	7	7	0	1
S2	B2	1	2	1	$2\overline{D3}$	0	1	1	0	0
S14	B2	1	0	1	$5\overline{D3}$	0	1	1	0	0

S17 B2	1 $\overline{D3}$	0	1	1	0	0	0
S28 B2	1	4 $\overline{D3}$	1	1	0	0	1
Remainder 3B2	3	3	3	3	0	0	0
S4 B3	1	1 $\overline{D3}$	1 $\frac{B3}{e}$	1	0	1	0
S11 B3	1 $\overline{D3}$	1	3	1	0	0	0
S29 B3	1	2	0	1	0	0	3 $\overline{B1}$
S34 B3	1	1	4 $\frac{B3\ D3}{B3}$	u	1	0	0
Remainder 5B3	5	4	6	3	0	3	0
TOTALS:							
N 33	33	41	63	30	33	3	8
Records Diagnosis B or D 33	1	2	7	1	0	0	0
Interview Diagnosis B or D —	5	5	11	0	1	0	2
Either Diagnosis B or D 33	5	6	14	1	1	0	2

[a] Number of relatives identified
[b] Remainder = probands with no schizophrenia diagnosis in either family
[c] Diagnoses above the line are based on institutional records; those below the line are based on the interview.
 d = dead; e = emigrated; r = refused interview; u = unknown

Table 4b. Consensus Diagnoses Based upon Institutional Records or Interviews of Schizophrenia (Definite: Chronic B1, Acute B2, Latent B3, or Uncertain: Chronic D1, Acute D2, Latent D3) in the Biological and Adoptive Families of Control Probands

Proband	Diagnosis	Biological family					Adoptive family			
		Mother	Maternal half-sibs	Father	Paternal half-sibs	Full sibs	Mother	Father	Half-sibs	Full sibs
Screened controls										
C6	A01*	1	0	1	3	0	1	1	0	$1\frac{D3}{r}$
C3	A41*	1	1	1	1	0	1	1	0	$1\frac{B3}{D3}$
C9	A55*	1	0	$1\frac{B1}{d}$	0	$1\frac{B3}{B3}$	1	1	0	0
Remainder	20A	20	21	19	38	3	20	20	0	16
Totals:										
N	23	23	22	22	42	4	23	23	0	18
Records diagnosis B or D	0	0	0	1	0	1	0	0	0	2
Interview diagnosis B or D	0	0	0	0	0	1	0	0	0	1
Either diagnosis B or D	0	0	0	1	0	1	0	0	0	2
Uncertain controls										
C4	r	1	$6\overline{D3}$	1	2	0	1	1	0	0
C20	r	1	0	1	1	0	$1\overline{D3}$	1	0	0
C18	e	$1\overline{B3}$	3	$1\overline{D3}$	$6\overline{D3}$	0	1	1	0	0
C21	e	1	1	1	$3\overline{B3}$	0	$1\overline{D3}$	1	0	$1\frac{B1}{B1}$

C28	e	1	2	1 $\frac{D3}{d}$	1	0	1	1	0	1
C26	d	1	0	1	1	0	1	1	0	0
Remainder	3C*,1e,1d	4	6	4	8	1	5	5	0	3

Totals:

N	11	10	18	10	22	1	11	11	0	5
Records diagnosis B or D	0	0	0	1	0	0	0	0	0	1
Interview diagnosis B or D	0	1	1	1	2	0	2	1	0	1
Either diagnosis B or D	0	1	1	2	2	0	2	1	0	1

Totals all controls:

N	34	33	40	32	64	5	34	34	0	23
Records diagnosis B or D	0	0	0	2	0	1	0	0	0	3
Interview diagnosis B or D	0	1	1	1	2	1	2	1	0	2
Either diagnosis B or D	0	1	1	3	2	1	2	1	0	3

*Diagnoses other than schizophrenia: A01–normal; A41–mild hysterical personality;
A55–severe hysterical personality
C–inadequate or schizoid personality

Table 5. Schizophrenic Illness in the Biological Paternal Half-siblings of
Schizophrenic Index Cases and Controls (as Obtained by Consensus
Diagnoses Based upon Institutional Records or Interviews)

Probands (N)	Number of biological paternal half-sibs	Number with diagnosis of schizophrenia					
		Definite		Uncertain		Total	
		N	(%)	N	(%)	N	(%)
Schizophrenic Index (33)	63	8	(12.7)	6	(9.5)	14	(22.2)
Control (34)	64	1	(1.6)	1	(1.6)	2	(3.1)
p (index vs. control)		0.015		0.055		0.001	
Screened controls (23)	42	0	(0)	0	(0)	0	(0)
p (index vs. screened control)		0.014		0.042		0.0004	

view any hunch which he had and its sources, which enabled him to venture a guess that this was a biological, adoptive, index or control relative. These items were also deleted before we made our independent diagnoses, but after that task was done, it was possible to examine the unedited transcripts and analyze Dr. Jacobsen's hunches and the information on which they were based.

In more than half the interviews, Dr. Jacobsen had no clues at all, and no hunches. In another 30 percent of the cases, he did not guess correctly, but in 58 interviews, or 17 percent, he surmised correctly the type of the relative. However, on reading the basis of his hunch, one makes an interesting discovery: that his clinical description was not influenced by his hunch but quite the other way around. He would reason like this: "this person looks very much like a schizophrenic," or "this person has a lot of the psychopathology of schizophrenia, and I believe I interviewed his half-sister, and she was also schizoid. I would imagine that these people are in the biological index group of relatives." Actually, his impression of psychopathology permitted a surmise regarding the type of relative. In any case, even disregarding the finding that his hunches usually did not affect his diagnosis, we can delete those cases in which his hunches were correct. Table 7 shows the distribution of consensus diagnoses of schizophrenia, chronic or borderline, definite or uncertain, based upon his edited interviews. (Consensus diagnoses were made by D.R., P.H.W., and S.S.K. without knowledge of Jacobsen's diagnoses, information, or hunches of relationships.) If we now delete from that tabulation those in which Jacobsen correctly surmised the type of relative, the highly significant con-

Table 6. Schizophrenic Illness in the Biological and Adoptive Families of
Schizophrenic Index Cases and Controls (Ascertained by Consensus
Diagnoses Based upon Institutional Records or Interviews)

Probands	Number	Number of families with one or more members diagnosed as schizophrenic							
		Biological				Adoptive			
		Definite		Definite or uncertain		Definite		Definite or uncertain	
		N	%	N	%	N	%	N	%
Schizophrenic Index	33	14	(42.4)	17	(51.5)	1	(3.0)	3	(9.0)
Control	34	3	(8.8)	5	(14.7)	3	(8.8)	5	(14.7)
p index vs. control		0.002		0.001		n.s.		n.s.	
Screened control	23	1	(4.3)	1	(4.3)	1	(4.3)	2	(8.7)
p index vs. screened control		0.001		0.0001		n.s.		n.s.	

centration of schizophrenia in the biological relatives of the index cases
remains relatively undiminished.

These results, based now on psychiatric interviews on relatives outside
of psychiatric institutions confirm results previously obtained only from
institutional records. The greater yield of psychiatric diagnoses made possible by the interview, however, permits a greater resolution into more
specific diagnostic categories within what was designated as the
"schizophrenia spectrum" in the earlier study. Not only is there a highly
significant concentration of diagnoses over that spectrum in the biological relatives of adoptees who became schizophrenic, but diagnoses of
chronic, latent, and uncertain schizophrenia are also significantly concentrated in that population and randomly distributed in the other three
populations of relatives (biological relatives of controls, adoptive relatives
of index cases, and of controls) none of whom are genetically related to the
schizophrenic index cases. This is strongly suggestive of the operation of
genetic factors, a conclusion which is confirmed by the concentration of
schizophrenia in the biological paternal half-siblings of the schizophrenic
index cases with whom they did not share *in utero* or neonatal experiences but only a certain amount of genetic overlap.

The data analyzed thus far do not pertain to possible environmental
factors which may operate in the development of schizophrenia, although
it is clear from these results that one of them—schizophrenic illness in the
rearing family—is not necessary.

Table 7. Effect of Possible Interviewer Bias on Distribution of Consensus
Diagnosis

Type of relative	Consensus diagnosis of schizophrenia: chronic or borderline, definite or uncertain	Number in which interviewer correctly surmised type of relative	Remainder without correct surmise
Biological index	24	5	19
Biological control	6	0	6
p	0.00005	0.03	0.001
Adoptive index	2	1	1
Adoptive control	5	1	4
p	n.s.	n.s.	n.s.

These findings have an unexpected but important bearing on another hypothesis which this research was not originally designed to test, since it has only recently been stated seriously. That is the notion that schizophrenic illness is a myth. These data do not permit the conclusion that schizophrenia is a unitary disorder, since they are equally compatible with a syndrome of multiple etiologies and different modes of genetic transmission. They make it difficult, however, to deny the existence of a syndrome which independent raters can agree upon quite reliably and which each rater without knowledge of the relationships finds significantly concentrated only in the biological relatives of schizophrenics. If schizophrenia is a myth, it is a myth with a substantial genetic component.

Diabetes mellitus is analogous to schizophrenia in many ways.* Both are symptom-clusters or syndromes, one described by somatic and biochemical abnormalities, the other by psychological. Each may have many etiologies and shows a range of intensity from severe and debilitating to latent or borderline, and for both there is evidence that genetic and environmental influences operate in their development. There appears to be little basis for regarding one as more a myth or less an illness than the other.

References

Heston, L. L. Psychiatric disorders in foster home reared children of schizophrenic mothers. *British Journal of Psychiatry*, 1966, **112**, 819.

Kety, S. S. Biochemical theories of schizophrenia. *Science*, 1959, **129**, 1528 and 1590.

*For a fuller discussion of the analogy with diabetes mellitus see the paper by Shields, Heston, and Gottesman in this volume.

Kety, S. S., Rosenthal, D., Wender, P. H., & Schulsinger, F. The types and prevalence of mental illness in the biological and adoptive families of adopted schizophrenics. In D. Rosenthal & S. S. Kety (Eds.), *The transmission of schizophrenia.* Oxford: Pergamon, 1968, pp. 345–362.

Rosenthal, D. Problems of sampling and diagnosis in the major twin studies of schizophrenia. *Journal of Psychiatric Research,* 1962, **1**, 116.

Rosenthal, D., Wender, P. H., Kety, S. S., Schulsinger, F., Welner, J., & Jacobsen, B. Evidence for a spectrum of schizophrenic disorders, 1974, to be published.

Rosenthal, D., Wender, P. H., Kety, S. S., Schulsinger, F., Welner, J., & Østergaard, L. Schizophrenics' offspring reared in adoptive homes. In D. Rosenthal & S. S. Kety (Eds.), *The transmission of schizophrenia.* Oxford: Pergamon, 1968, pp. 377–391.

Wender, P. H., Rosenthal, D., & Kety, S. S.: A psychiatric assessment of the adoptive parents of schizophrenics. In D. Rosenthal & S. S. Kety (Eds.), *The transmission of schizophrenia.* Oxford: Pergamon, 1968, pp. 235–350.

12 SCHIZOPHRENIA AND THE SCHIZOID: THE PROBLEM FOR GENETIC ANALYSIS

James Shields, Leonard L. Heston,
and Irving I. Gottesman

Recent studies of families of schizophrenics have uncovered a problem that has lain dormant for decades—how to categorize the large proportion of relatives of schizophrenics who are afflicted with psychopathology other than typical schizophrenia. The problem is important to those concerned with genetics, because genetic analysis requires a definable trait. It is also important for many other reasons. The investigator with a trait apparently specific to schizophrenia might wrongly interpret his results or even wrongly discard them if he found the same trait in nonschizophrenic first-degree relatives of schizophrenics. If a control group for schizophrenia research is wanted, the investigator must be aware that psychiatric and clinic populations are likely to contain significant proportions of behaviorally impaired but nonschizophrenic persons who possess genes associated with phenotypic schizophrenia. Clinicians, including genetic counselors, need as much data as can be made available about families of schizophrenics.

Attempts at describing and then devising theories accounting for the abnormal relatives of schizophrenics, or "schizoids" as we shall call them, go back at least to Gadelius, who in 1909 (according to Essen-Möller 1946) remarked on the unreasonableness and inaccessibility to argument he found among relatives of schizophrenics. During the following decades the schizoid relatives of schizophrenics were prominent in the concerns of most major investigators and schools of psychiatry (review by Planansky, 1972). However, their theoretical formulations have been made largely obsolete by advances in genetics and, after making limited but notable progress, their descriptive efforts foundered on the same problems we face today. We have no exact solutions to the problems, but we believe that the great prog-

James Shields, B.A., Senior Lecturer, Institute of Psychiatry, London, England.

Leonard L. Heston, M.D., Professor, Department of Psychiatry, Research Unit, Medical School, University of Minnesota, Minneapolis, Minnesota.

Irving I. Gottesman, Ph.D., Professor, Departments of Psychology and Psychiatry, University of Minnesota, Minneapolis, Minnesota; Guggenheim Fellow (1972–73), Psykologisk Institut, Copenhagen, Denmark.

ress made in genetics, coupled with new data from recent family and twin studies, adds substantially to our understanding. However, before taking up those topics we must first clear away the logical and semantic debris that has plagued the concept of the schizoid from the beginning.

What Is Inherited in Schizophrenia?

In an obvious sense, it is not the schizophrenic psychosis per se that is inherited. The distinctive delusions and hallucinations, the first-rank symptoms of Kurt Schneider, the Bleulerian secondary symptoms—these all develop in the predisposed individual as the result of a series of ontogenetic interactions between his constitution and his environment. It is only DNA with its encoded sequence of nucleic acids which is inherited. All morphological, physiological, and behavioral traits are acquired in the course of development. But at our present level—where we must begin—those who are interested in the schizoid, or schizoidia as it is sometimes called, are not as a rule concerned with a molecular basis for schizophrenia, but with attempts to identify a behavioral trait, basic to at least some schizophrenias, which one could speak of as being inherited in the generally accepted sense in which, say, Huntington's chorea may be said to be inherited (despite the wide variation in age at onset, psychiatric symptomatology, and present ignorance of a biochemical basis), or in which stature may be said to be inherited (despite the role of nutrition and the many individually indistinguishable genes presumably involved). The underlying hope, which may or may not be justified (Shields 1971), is that a "schizoid" trait may prove a better phenotype for genetic analysis than schizophrenia itself.

Semantic Problem of the Schizoid

The main source of confusion regarding the term "schizoid" is the extent to which it denotes a psychological resemblance to schizophrenia or a genealogical connection with it, or, as Essen-Möller (1946) believed, both. At least four uses of schizoid may be distinguished.
1) Schizoid in the literal sense of resembling schizophrenia. This is how the word is used in the accepted diagnostic term *schizoid personality*, meaning shy, sensitive, aloof, or eccentric (American Psychiatric Association 1968; General Register Office 1968). It does not imply a genealogical or etiological connection with schizophrenia. It shades into the normal. It can be extended to include paranoid personality. Though not standard usage, it could also be extended to cover persons such as those with a T score of over 70 on the Schizophrenia scale of the MMPI or who score highly on a test of thought disorder that differentiates schizophrenics from others. It would not include depressives, criminals, or the mentally retarded, since they cannot generally be described as schizophrenic-like.

2) Psychiatric disorders occurring in the families of schizophrenics (usually the twins or first-degree relatives), whether resembling schizophrenia or not, and whether or not they occur more frequently in schizophrenic than control families. A genetic connection with schizophrenia is not implied. (These are the potential or eligible components for Heston's "schizoid disease.")

3) Disorders, whether occurring in a person who is the relative of a schizophrenic or not, that belong to a class found more often in the families of schizophrenics than controls. (These are akin to the "schizophrenia spectrum disorders" of Rosenthal and Kety's group, including those of the "extended" spectrum [see p. 183].)

4) A diagnosis or behavioral trait or combination of traits, whether diagnosable as abnormal or not, which is believed to indicate either a probable carrier of the schizophrenic gene (monogenic hypothesis) or a high-risk genotype (polygenic hypothesis). This usage resembles the schizotype of Rado (1962) and Meehl (1962, 1973), including the compensated schizotype, but is not necessarily wedded to Meehl's (1964) checklist of schizotypic signs or to a monogenic hypothesis.

Rather than arbitrarily modify the meaning of terms used by the DSM II, Heston, Rosenthal, or Meehl, we shall refer to these uses of the term schizoid as Sd 1, Sd 2, Sd 3, and Sd 4, respectively. Sd 1, 2, and 3 overlap to an extent that cannot be determined until much more extensive epidemiological and family investigation has been carried out on Sd 1 and Sd 3 cases.

We shall illustrate the overlap of Sd 1, 2, and 3 schematically. Let us start with persons in a population who have Sd 3 conditions. They are represented in Figure 1 by a circle with a dotted outline. These, then, are people with disorders of a kind found to occur more frequently in the relatives of schizophrenics than in controls. According to Kety, Rosenthal, Wender, and Schulsinger (1968) they would include criminals and according to Heston (1966) some mentally retarded. Of course, which disorders are identified as Sd 3 will differ from study to study, and their prevalence will differ from population to population. How far they may be related to schizophrenia genetically (Sd 4) is another matter.

Sd 1, according to our terminology, means 'resembling schizophrenia'. While some Sd 1 conditions, such as schizoid psychopathy according to the earlier investigators, almost certainly belong to Sd 3, other schizophrenic-like conditions may not necessarily distinguish schizophrenics' relatives from controls. Overinclusive thinking is an example; and it is unlikely that being *un*married (which some might conceivably call a schizophrenic-like condition) would be found in many studies to be a statistically significant Sd 3 trait. Some overlap between Sd 3 and Sd 1 is illustrated by the overlapping circles of the figure (Figure 1b) in which Sd 1 persons are represented by the vertically hatched circle. The vertically hatched area within the Sd 3 circle brings out the point that only some Sd 3 conditions resemble schizophrenia clinically.

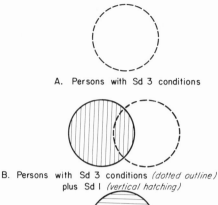

A. Persons with Sd 3 conditions

B. Persons with Sd 3 conditions *(dotted outline)*
plus Sd I *(vertical hatching)*

C. Persons with Sd 3 conditions *(dotted outline)*
plus Sd I *(vertical hatching)* plus Sd 2 *(horizontal hatchings)*

Fig. 1. Schematic Relationships among Sd 1, 2, and 3.

We shall now add another circle to represent psychiatrically abnormal persons (other than schizophrenics) found among the relatives of schizophrenics, that is, what we are calling Sd 2. Some of these persons will suffer from Sd 1 or Sd 3 disorders, but others will have conditions, such as anxiety states, which neither resemble schizophrenia clinically nor are generally found significantly more often than in a control group. They are shown by the horizontally hatched circle in Figure 1c which extends beyond the other two circles. The figure also shows that there may be many Sd 1 and Sd 3 individuals who are not closely related to a schizophrenic.

It has been suggested that people classed as schizoid according to each of these definitions might provide a better phenotype for genetic analysis than does a schizophrenic psychosis, and they might give an indication of what is inherited. However, with each the question arises as to how homogeneous the group is genetically. Are all criminals (Sd 3), or all over-inclusive thinkers (Sd 1), or all anxiety neurotics related to a schizophrenic (Sd 2), predicted to be high-risk candidates for developing and transmitting schizophrenia; and if not, what proportion of them are hypothesized as Sd 4, that is, schizoid in the genetic sense? Opinions on these points differ and are not well formulated. The search with which we and others are concerned is for an identifiable characteristic, whether dimensional or categorical in form, which is the best attainable indicator of Sd 4 in a defined population. While one would probably look for such a characteristic within Sd 1, 2, or 3, we may note that all schizophrenic genotypes do not reveal themselves easily; not all schizophrenics have a schizophrenic relative, nor are they all schizoid premorbidly in either the Sd 1 or Sd 3 sense.

It is obvious that considerable caution is required before claiming to have identified a phenotype which can be substituted for schizophrenia in population genetic studies. However, persons who are schizoid in any of these senses may provide promising leads toward a better understanding of the development of schizophrenia at a biological or any other level. In particular, the relatives of schizophrenics remain a strategic population.

Circular Argument or High-Risk Methodology?

Having outlined the semantic problem, we now turn to a logical one. As we have seen, some theorists consider certain traits as genetically related to schizophrenia (i.e., as part of Sd 4) only when they occur in schizophrenics' families (i.e., as Sd 2). Relatedness to a schizophrenic is not a very satisfactory concept for population genetics. The likelihood of having a first- or second-degree relative who is a schizophrenic depends, among other things, on the size and age distribution of the family, and this differs widely in ways that are difficult to allow for, for instance as between a parent born in 1900 and a child born in 1960. It would be hard to say how many persons in a given population have a schizophrenic relative. Furthermore, while a theorist may wish to predict that a trait indicates Sd 4 only when Sd 2, he cannot reasonably claim that no trait-bearer who is not Sd 2 can be Sd 4, and he has difficulty in specifying how many of them there might be. Most schizophrenics, after all, though presumed to be Sd 4, are not Sd 2, having no close relatives who are schizophrenic.

More serious to many minds, however, are what appear to be the logically circular elements in the argument. According to Mosher, Stabenau, and Pollin (1973) Heston (1970) argued as follows:

> We will demonstrate the genetic factor in schizoidia by defining a certain pattern of distribution of schizoid characteristics in the relatives of schizophrenics; and we will know that a given clinical picture is to be called schizoid if it is found in a subject showing the hypothesized relationship to a schizophrenic. Indeed, the American Psychiatric Association criteria for classifying schizoid personalities are rejected by Heston because they are not regularly found in the relatives of schizophrenics and do not meet his criteria of genealogic consistency. The fact that Heston himself points out the circularity of his type of analysis does not lessen the fact that it is, on the face of it, totally inadequate and unacceptable. Given such an approach, there is no way to disprove any hypothesis.

The criticism misunderstands the aim of the hypothesis. Heston's argument should be seen not as the *demonstration* of a genetic basis of schizoidia but as a *hypothesis* about the genetic aspects of schizophrenia. (That there are such aspects worth investigating is well attested by other evidence.) It is akin to hypotheses about identifying carriers in genetic disorders and may be looked upon as part of high-risk methodology studies in general. The theory is open to testing. It predicts that in pedigree data schizophrenia (narrowly defined) will be transmitted only through persons who suffer from schizophrenia or other functional disorders ("schizoid disease"). Given the generally accepted assumption that a given trait can have different causes, it is not question begging to suggest that certain traits occurring in schizophrenic families will generally bear a genetic relationship to schizophrenia, while in other groups such a relationship will be rare. As an imaginary illustration, let us assume that several children of parents with Huntington's disease are observed to be fidgety and that it is hypothesized that they are the carriers of the abnormal gene. It is predicted that the fidgety children *in Huntington families* will develop Huntington's chorea and that their siblings will not, but it is not predicted that all fidgety children in the general population are at risk for Huntington's disease. To the extent that fidgetiness is a common characteristic of children, the indicator will be an imperfect one. Some children of Huntington patients may be fidgety for reasons other than the specific Huntington gene. But the trait could nevertheless be a good, if imperfect, prognostic index in these families. It is not a 'totally inadequate and unacceptable' approach. Ultimately one would hope to discover an indicator closer to the site of action of the gene which would give no false negatives in Huntington families and no false positives in any family. The point as regards schizophrenia is illustrated in Figure 2. The situation is more complex and obscure in schizophrenia, and Heston's hypothesis may not be correct. But if any such hypothesis were to succeed in introducing greater orderliness to the family data it would have considerable merit. It makes sense to us to see whether a high-risk hypothesis, genetic or environmental, works in schizophrenic

Fig. 2. Phenotypic Sd 4 Traits (*dashed line*) as Indicator of Genotypic Sd 4 (*solid line*).

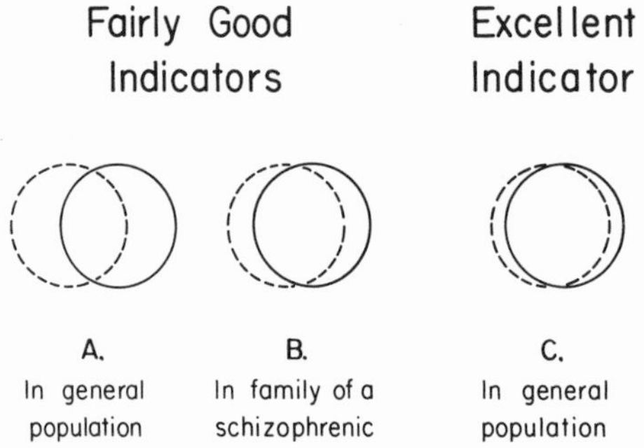

Fairly Good Indicators		Excellent Indicator
A.	B.	C.
In general population	In family of a schizophrenic	In general population

families without arguing that it must work equally well in the general population.

It is naive to object, as one frequently hears done, to selection of one of several data-classifications on the empirical basis that it yields greater genetic orderliness. Psychologists and psychiatrists who make this criticism usually display an undergraduate comprehension of philosophy of science. There is nothing viciously "circular" about an argument that begins with an observational-statistical finding, to wit, that one way of classifying events or entities leads to markedly greater empirical lawfulness than another way, and infers therefrom that the first classification is "better." *Per contra*, the most general characterization of scientific inference—cutting across all sciences and certainly not peculiar to problems in behavior genetics—is that we seek to maximize the order in phenomena. The bat "goes with" the whale despite the superficial and more common-sense grouping of whale with pickerel, and bat with bird. (Meehl 1972, pp. 385-86.)

There is no more unwarranted circularity in using the characteristics of the relatives of known schizophrenics to generate genetic hypotheses than environmental ones, for example, by trying to elucidate "transactions" in schizophrenic families by improved methods of scoring tests given to schizophrenics and their families (Wynne 1968) or by studying family interaction in discordant pairs of twins (Mosher, Pollin, & Stabenau 1971). How far conclusions about the genetic basis of other psychiatric disorders or the relevance of identification or birth weight or submissiveness in twins apply outside "schizophrenic" families is a matter for further investigation after they are proved relevant within such families.

Variety of Methods and Assumptions

Attempts to improve our understanding of what is inherited in schizophrenia have come from studies on various relatives of schizophrenics: on parents and on children (including children reared away from their biological parents), on monozygotic twins, on siblings, and on half-sibs. These have been compared with other relatives of schizophrenics (e.g., MZ with DZ twins), with the relatives of probands from other diagnostic groups (e.g., neurotics), with restricted groups unrelated to a schizophrenic but appropriately matched, say, for adoptive or foster-home status and social class, and with findings reported for the general population. Studies have come from countries as far apart as Finland, Japan, and the United States. Within the same country (Denmark) there have been recent studies of twins born 1870-1920 and of adoptees born 1924-47. Schizophrenia has sometimes meant process schizophrenia; sometimes it has included possible borderline schizophrenia. Some studies have been based on records only, others on intensive interviews and extensive fieldwork. Precautions have not always been taken to avoid diagnostic contamination. Background assumptions have differed widely about the possible heterogeneity of schizophrenia, about the relationship of schizophrenia and other disorders, about the time of life when the genes exert their most distinctive effect in schizophrenia, and about the role of the environment. It is therefore hardly surprising that findings and interpretations have differed correspondingly widely as to what conditions might be genetically related to schizophrenia and in what way. One of us (Shields 1971) considered there might be almost as many opinions as investigators. That may be an understatement, since the same investigator has sometimes expressed more than one opinion.

We shall not attempt to review the whole field. We shall select for comment some of the recent studies with which we are most familiar. We shall note what disorders are found in the relatives studied (Sd 2) and ask how far they are schizophrenic-like (Sd 1) and differ in frequency from control groups (Sd 3). Assuming that some of the disorders (Sd 4) indicate a schizophrenic genotype, do they fit a genetic hypothesis? We shall outline the more plausible hypotheses as they arise in the context of a study under discussion.

Schizoid Psychopathy in Recent Family Studies

Alanen. First, a family study from Finland. Alanen (1966) was concerned with the psychopathogenesis of schizophrenia. He compared thirty schizophrenic with thirty neurotic families. The investigation of relatives was one of the most intensive ever undertaken in schizophrenia. Perhaps because of this or because of a greater propensity to find psychopathology, more diagnosable afflictions were found by Alanen than by most other investigators. While investigator bias cannot be excluded, Alanen's perspective was psychodynamic rather than genetic, and ratings by a psychologist who did not know the family membership of the subjects are available for comparison. Our concern is with the nature and extent of abnormality present in the parents and siblings. The results of Alanen's study can be seen in Table 1.

The salient feature of Table 1 is the striking excess of psychotic and personality disorders among the relatives of schizophrenics as compared with relatives of neurotics. This division of Alanen's data shows only the more severe cases, that is, those receiving a global psychopathology rating of IV or more on Alanen's scale, which ran from I to VI. Persons scoring I, II, or III are counted as normal in our table. The difference between the groups of relatives remains, no matter what level of severity is used.

Judging from the case history material in Alanen's monograph, his diagnosis "schizoid character" would correspond with standard practice in the U.S. and U.K. Hence schizophrenia + Sd1 in his groups is relatively clearly defined by the diagnoses numbered by us in the table as 1–3. Sd 2 is also clearly set out; it is items 1–18 on the left half of Table 1—that is, if we are prepared to omit Alanen's ratings II and III because they are nearly normal. Sd 1 + Sd 3 seem to correspond with what the earlier European investigators might have described as schizophrenia, together with schizoid and paranoid psychopathy. While it may be reasonable to nominate most Sd 3 diagnoses as candidates for Sd 4 (i.e., a genetic connection with schizophrenia) in Alanen's study, it is more problematical to assess how many of the relatives of schizophrenics among those with diagnoses 12–19 (character neurosis to nearly normal) might also be Sd 4; some surely will be, and one might propose those who, at this level of severity, can still be described as having schizophrenic-like features.

No simple genetic hypothesis can explain these data. A global predisposition to nonspecific psychopathology associated with a polygenic base (a theory closest to Alanen's own) is not compatible with these data; there is an evident discontinuity between the relatives of neurotics and those of schizophrenics. An autosomal dominant hypothesis greatly modified to account for variability in expression (some would say penetrance rather than expression) is tenable. So is a polygenic hypothesis which, to be realistic, must be modified by positing thresholds and by making provision for genes with unequal effects in a polygenic system. In our opinion, the dominant main gene and the polygenic models are the bases for the only hypotheses concerning the bulk of schizophrenia that are compatible with the present available evidence. Indeed, we think there may well be elements of both in the final accounting. Some allowance will, of course, have to be made for etiological heterogeneity.

Bleuler. Another recent family study with careful personal investigation is Manfred Bleuler's (1972) longitudinal observation of 208 Burghölzli schizophrenics (his own patients) and their families over a period of twenty-two years. Bleuler can claim a deep insight into the clinical nature of the schizoid, and has an unrivaled knowledge of the literature. Earlier studies have left him in no doubt that there is a marked excess of schizoid psychopathy (Sd 1) among parents and other relatives of schizophrenics. In a previous study of his own he had reported a rate of 16.5 percent among parents of schizophrenics, compared with 1.2 percent among parents of patients with physical illnesses. He did not feel it necessary to demonstrate the fact once again in his present study. This was partly because the longer he knew his families, the more difficult he found it to decide whether a member could be called a schizoid psychopath—there could be no exact figure—and partly because such a diagnosis might be taken to imply a fixed constitutional trait. Bleuler was impressed by the environmental contribution to the schizoid picture. We agree with him that personality traits as well as psychosis are influenced by the environment. Nine percent of the children of the 208 schizophrenics were also schizophrenic, a rate similar to that found in most other studies, but only a further 5.6 percent could be

Table 1. Disorders in Families of Neurotics and Schizophrenics (after Alanen 1966)

Diagnoses	Families of schizophrenics				Families of neurotics			
	Fathers	Mothers	Brothers	Sisters	Fathers	Mothers	Brothers	Sisters
1. Schizophrenia	1	2	3	1				
2. Borderline schizophrenia	1		1	1				
3. Schizoid character	1	10	2	1				
4. Paranoid psychosis	2	1						
5. Jealousy paranoia	2	1						
6. Paranoid psychopath	2							
7. Paranoid character	1	1	1					
8. Depressive psychosis	1							
9. Psychotic personality		1						

10. Borderline personality		1		2				
11. Primitive character		1						
12. Character neurosis	4		1	1			1	
13. Alcoholic psychopath	3	2			3		1	
14. Alcoholic	2							
15. Psychopath	1			1	1		1	
16. Neurosis		1	1					
17. Character disorder	1	1		1		4	1	
18. Personality disorder		1		1				
19. Normal or nearly normal*	9	10	16	13	24	25	23	23
Totals	30	30	28	21	30	30	27	23

*Includes all subjects given a rating of III or less on Alanen's scale of psychopathology which ranged from I to VI with increasing psychopathology.
III = Psychoneurosis, psychoneurotic personality.
IV = Disorders more severe than psychoneurosis, which includes some complicated neuroses, e.g., "neurosis with borderline schizophrenic features."

termed schizoid. Despite the environmental circumstances, as many as 72 percent of the children were normal. Of the 5 children both of whose parents were schizophrenic, 2 were schizoid and one severely neurotic, but none was schizophrenic.

In view of what follows, it may be of interest to note that Bleuler found the previous personality of schizophrenics to be morbidly schizoid in 24 percent of cases, and schizoid but within normal limits in 30 percent; the personality was conspicuous in other ways (abnormal or within normal limits) in 16 percent, and unremarkably normal (*unauffällig*) in 30 percent. About 7 percent of the parents were schizophrenic or probably so; there may also have been a slight excess of unipolar affective psychoses and senile psychoses compared with general population rates. As many as 24 percent of the fathers were alcoholic, but, unexpected though it may seem, this was considered to be little more than the then current prevalence in Zurich. Bleuler would be reluctant to count alcoholism as Sd 3, let alone Sd 4. For full sibs the schizophrenia risk was about 10 percent; there was no excess of other psychoses, but a variety of personality disorders was noted. As in other studies, the illnesses of pairs of schizophrenic sibs supported the view that different forms of schizophrenia are related, though there was a tendency toward familial resemblance, particularly with respect to recurrent schizophrenia with good prognosis. Among 258 half-sibs the schizophrenia risk was 5.3 percent; other mental disturbances were rare.

Lindelius (1970). This study of schizophrenia from rural south Sweden does not deal specifically with the schizoid (Sd 1), but it is relevant to the question whether other abnormalities in the families (Sd 2) might be Sd 4. Like Kallmann's (1938) large Berlin study, it was based on probands hospitalized in the early part of the century, permitting lengthy observation of relatives. The risks for schizophrenia in the sibs and children of the 270 probands are consistent with the findings of Bleuler and most other workers—9.1 percent and 9.5 percent respectively—when adjustment was made for "undiagnosed" psychosis. As in Hallgren and Sjögren's (1959) Swedish study the rate in parents was very low (1.7 percent)* but, as in several other studies, many were alcoholic (16 percent) or psychopathic (7 percent) or committed suicide (8 fathers and 6 mothers). Some of these abnormalities could arguably have been genetically related to the schizophrenia of the proband. The findings with respect to nieces and nephews suggest more clearly that some of the "other disorders" were genetically related to schizophrenia in those families. When the parent of a nephew or niece was disordered in any way, the schizophrenia rate rose to the surprisingly high figure of 12.0 percent ± 2.1; when the parents ap-

---

*It is difficult to diagnose certain schizophrenia in parents of patients first hospitalized in 1900.

peared healthy, it was 2.6 percent ± 0.5. Kallmann (1938) found much the same thing.

"Spectrum Disorders" in Fostering Studies

Heston. We now turn to first-degree relatives as described in some adoptive/foster-rearing studies. Heston (1966) reported a detailed investigation of 47 persons born to schizophrenic mothers in Oregon state hospitals compared to 50 matched controls born to mothers with no known history of psychiatric disorder. Both groups were reared in adoptive or foster homes. Table 2 presents the results. One can analyze these data in the same way as was done for Alanen. The diagnoses, being based on different nomenclatures, differ, and Alanen found somewhat more psychopathology among parents and sibs than Heston found among children. Eight of the nine antisocial personalities in the experimental group and neither of the two in the control group fitted the older diagnostic category of "schizoid psychopath" (Sd 1), which is in line with Alanen's findings.

One diagnostic category, mental retardation, warrants some added attention. Retardation has been found associated with schizophrenia in several studies. Kallmann found about 10 percent retarded among children of simple schizophrenics. Hallgren and Sjögren (1959) surprisingly (c.f., Bleuler 1972) found several of their schizophrenics to be retarded, but there was no excess of retardation in their families. Retardation is not a part of the Rosenthal–Kety spectrum, and the weight of opinion has been against an association between retardation and schizophrenia. Also, there is evidence from Kallmann and from Brugger's (1928) study on *Pfropfschizophrenie*, where the two disorders, schizophrenia and mental retardation, segregated independently. It now appears that the disorders do segregate independently, but there is, nevertheless, an association. Jones (1973) points out that schizophrenics of lower intelligence tend to have earlier onset, more severe disease, spend longer time in hospital, and have fewer children than schizophrenics of higher IQ. Also, schizophrenics, on the average, have lower IQ's

Table 2. Adopted and Fostered Children of Schizophrenics and Controls
(Heston 1966)

	Experimental	Control
Number	47	50
Schizophrenia	5	0
Antisocial personality	9	2
Neurotic personality	13	7
Mental deficiency (IQ < 70)*	4	0

*One mental defective was also schizophrenic.

than their nonschizophrenic siblings. The evidence thus points to IQ as being one factor that tends to modify the expression of schizophrenia. Less intelligent schizophrenics spending longer periods in hospital would be more likely to be included in prevalence studies. They would have been more likely to be included as probands in studies such as Kallmann's and Heston's than in Rosenthal's or Kety's. Lower intelligence in the probands makes it more likely that the milder familial sorts of retardates would appear among relatives, as was in fact observed. Thus the association is a real one, but retardation is not a schizoid disorder in the sense of Sd 1 (as was obvious), Sd 3, or Sd 4.

Rosenthal et al. Rosenthal, Wender, Kety, Schulsinger, Welner, and Østergaard (1968) compared Danish adoptees of psychotic (or borderline psychotic) and normal parentage. As in other studies by this group, every precaution was taken to avoid diagnostic contamination. With findings in a few more cases outstanding, the latest provisional report (Rosenthal, Wender, Kety, Welner, & Schulsinger 1971; Rosenthal 1971, 1972) accounts for 76 index children and 67 controls. We have tabulated the results in Table 3. The grouping of the diagnoses is ours, as is the classification of the children into the groups employed; difficulty arose in only a few cases. The symbols, B_1, etc., are those used by Rosenthal et al. to classify the parents:

B_1 = process schizophrenia, diagnosed by all judges;

B_1/ = process schizophrenia, diagnosed by at least one judge but not by all judges; B_2 = reactive schizophrenia; B_3 = borderline or pseudoneurotic schizophrenia; M = manic-depressive psychosis; B_2/M = judges could not agree which of these two diagnoses was correct; B/D/M/A = judges could not agree which of these diagnoses applied to a particular case (A = not in the schizophrenia spectrum; D = doubtful schizophrenia). The authors originally (Rosenthal et al. 1968, p. 387) said their "data suggest that the inherited core diathesis is the same for both schizophrenia and manic-depressive psychosis." The size of the diagnostic subgroups in this unique sample is too small to permit firm conclusions. However, the breakdown in Table 3 equally suggests that inheritance in schizophrenia, or at least in process schizophrenia, may be more specific. Unqualified, albeit mild, schizophrenia occurred only in the offspring of process or possibly process schizophrenics. The only manic-depressive offspring had a manic-depressive parent. Borderline schizophrenia, when not qualified in some way, shows up on this analysis as an Sd 3 characteristic (and also presumably Sd 4), but characteristics lower down the "spectrum," such as qualified "borderline" conditions, or even schizoid personality (Sd 1), do not. "Spectrum" diagnoses were made in as many as 12 out of 67 control adoptees (18 percent) in this study, remarkably high in our opinion.

Although the results of Heston's and Rosenthal's studies of the offspring of schizophrenics raised apart from their biological parents agree in the

Table 3. Diagnosis of Adoptee Children by Diagnosis of Biological Parent (Data of Rosenthal 1971)

Adoptee	Index biological parents (76)				Control biological parents (67)
	Process or ? process schizophrenia B1, B1/	Acute or borderline schizophrenia B2, B3	Possibility of manic-depressive psychosis B2/M,B,D/M/A	Manic-depressive psychosis M	Not psychotic
Schizophrenia, unqualified	3	–	–	–	–
Borderline schizophrenia, unqualified, or beginning schizophrenia?	4	–	1	–	3
Borderline schizophrenia, qualified[a]	5	1	2	1	5
Manic-depressive	–	–	–	1	–
Other spectrum diagnosis[b]	–	1	5	–	4
No spectrum diagnosis	32	6	9	5	55
Total	44	8	17	7	67

[a]E.g., Paranoid borderline; Close to borderline psychotic.
[b]E.g., Schizoid personality; Paranoid character.

181

main, there are differences in what they found to be Sd 3. Only a few of the differences and their possible origins will be illustrated. In addition to the absence of mental retardation in the Danish study (mentioned above), there was no sociopathy or criminality in the Rosenthal et al. Index adoptees; 8/30 males in Heston's experimentals were sociopathic personalities, were felons and had prison records, whereas none of Rosenthal's 35 males (Wender, Rosenthal, Kety, Schulsinger, & Welner 1973) had a criminal history. There are other relevant data. Hutchings (1972) worked with Mednick and Schulsinger on the same Adoptee Register used by Rosenthal et al. His findings, based on a careful search of police records in Denmark, were that 16.2 percent of male adoptees consecutively admitted to the register (1927–41) had a criminal record for other than minor offenses. In the equal sized control group of 1,145 nonadopted males the rate of criminality was 8.8 percent. The calculation of a 99 percent confidence interval for an expected number of Danish adoptees with a criminal record in Rosenthal's sample of 35 males shows that finding no criminals could not happen by chance. The question whether sociopathic disorders could be Sd 2 and Sd 3 conditions must be left open.

There are a number of methodological differences between the Rosenthal et al. and Heston studies which might account for the different findings. Several features of Heston's procedures for locating index children tended to bias his group of mothers toward chronic severe disease (Heston 1966). Most mothers were in hospital for life, and all would be unequivocal cases of dementia praecox. Being Oregon State Hospital patients, Heston's mothers were also biased toward lower social class. Rosenthal's schizophrenic parents included many milder cases and were probably more representative of all social classes. Social class differences between various strategic populations are ably discussed by Wender et al. (1973). Severity of illness in a proband case is associated with the presence of disorder and severity in their relatives (Kallmann 1938; Gottesman & Shields 1972). Low social class is associated with higher prevalences of psychiatric disorders. All of Heston's schizophrenic parents were mothers; Rosenthal's included 26 fathers. Sex of parent may be an important variable in accounting for severity in children. Heston did not systematically investigate his fathers. Although the putative fathers appeared to be mostly of the ordinary working class, and none were known to be hospital patients, the severe illness of the mothers, together with the possibility of illegitimacy and the low class of the families, was no doubt associated with a high level of nonspecific background psychopathology.

Although there is no direct evidence, two other differences may be important. Heston started with mothers who had borne children while patients in hospital. While not all were psychotic at the time of birth, all had been. Only 2 of the 50 Danish mothers were in a mental hospital while pregnant with the study child and only 10 of the 76 parents had their first psychiatric admission before the birth of the child. There may well have been some screening by the Danish adoption authorities to avoid the placement of children of psychotics, as Rosenthal (1971) suggested. This may be a critical point. The study group of Rosenthal et al. included only legally completed adoptions. Heston's included all children not reared by their mothers or maternal relatives; some of Heston's group, including all 4 retardates, were reared by paternal relatives or in child-care institutions supplemented by foster-family placements. Retarded children, and presumably others thought to be especially high-risk, would probably have been kept out of adoptive homes by adoption agencies, and hence out of Rosenthal's group. Finally, there may well be real gene pool differences or differences in the social milieu between Oregon and Denmark associated with differences in psychopathological expression.

We make these points because, as should now be obvious, we believe that no single approach to the problem of identifying schizotypes is free from intrinsic biases and that conflicting results are to be expected. Hence, although Heston and Rosenthal agree in the main as regards schizophrenia, we should not be surprised that there are differences in what we call Sd 2 and Sd 3.

Kety et al. (1968). More data from this study, a companion to that of Rosenthal et al. (1968), is presented in this volume, and we will therefore limit our comments.

A finding of interest is the apparently high rate of 'extended spectrum' cases, such as criminality and character disorder, found in the biological parents of adoptees who later became schizophrenic. The characteristics of parents who place their children for adoption need to be known before the general significance of such findings can be assessed. Hutchings (1972) found that 36.4 percent of 971 biological fathers of consecutively registered Danish male adoptees, unselected for any disorder or diagnosis, had a criminal record, compared with 11.4 per cent of 1,118 control fathers of nonadoptees. A history of psychiatric illness was found in 11.5 per cent of 286 biological parents of adoptees, compared with 5.6 per cent of 286 adopting parents. Adopting parents are a relatively healthy group compared to the general population, while parents yielding children for adoption are relatively unhealthy. Had Kety et al. not included control adoptees to make allowance for the sampling bias toward parental pathology, Hutchings' data might lead us to expect that these extended spectrum (Sd 2) parents might—like the alcoholic fathers in Bleuler's study—not be suffering from Sd 3 conditions. In view of the high base rate and low signal-to-noise ratio, it is all the more striking that only 2 such conditions were found in the 63 biological parents of control adoptees, compared with 9 in those of the schizophrenic adoptees. While one would not expect such a high rate in parents of schizophrenics who do not place their children for adoption, it is possible that some of these 'extended spectrum' parents are nevertheless Sd 4. It would be interesting to learn whether any of them are clinically Sd 1.

These elegant NIMH-Danish adoption studies employed appropriate controls for their purposes. Our observations are not intended as criticisms. Rather, they point out the sampling problems intrinsic to the adoption/foster-rearing strategy and, more specific to our purpose here, the difficulty of detecting the Sd 4 signal in the face of so much background noise.

Normality and Schizoid Traits in MZ Co-twins

We now turn to the evidence from twins. Provided we assume that non-genetic "symptomatic" schizophrenias are rare enough to be ignored and that MZ twins are genetically identical, the MZ co-twins of schizophrenics should be able to tell us something about the range of phenotypic manifestations (Sd 4) of schizophrenic genotypes, i.e., about the schizophrenia spectrum in one sense of the term. (Although random inactivation of the X chromosome in female pairs and possibly of parts of autosomes in both male and female pairs may require some modification of the twin method, we will for now assume that all MZ twins of schizophrenics will have a schizophrenic genotype—are Sd 4.) In some studies the co-twins are as often neurotic as schizophrenic (narrowly diagnosed), and it is then sometimes implied (Kringlen 1967) that it is therefore only a tendency to mental disorder in general that is inherited. Do we then regard all neurotics as belonging to the schizophrenic spectrum? Clearly not. By the same argument, normality would be part of the spectrum too. In one study (Fischer, Harvald, & Hauge 1969) 43 percent of MZ co-twins were clinically normal, a higher percentage than in any of the three other main categories employed.

Table 4 shows the extent to which normality and nonschizophrenic disorders were diagnosed in 7 twin studies. Normality could be paired with severe as well as mild schizophrenia. Differences in the reported "normality" rates shown may depend partly on the extent of the investigation and standards adopted for what is within the normal range, but they probably also depend considerably on the varying use made by different authors of the ambiguous term schizoid and whether persons with a few schizoid traits were regarded as normal or not. In his original report Tienari (1963) described none of his 16 MZ co-twins as schizophrenic. Six had other psychiatric diagnoses and 10 were normal; 12 twins, many of them healthy, displayed schizoid traits. Essen-Möller's (1941) and Inouye's (1963) results are also difficult to tabulate. Both regarded all nonschizophrenic MZ co-twins as schizoid (Inouye) or as having a characterological trait genetically related to schizophrenia (Essen-Möller). Mosher et al. (1973) reported that only 6 out of 15 nonschizophrenic MZ co-twins were schizoid, either in the DSM II sense or in being rated highly in respect of the traits which

Table 4. Pairwise MZ Rates for Schizophrenia, Schizoid, and Other Psychiatric Conditions, and Normality in Some Schizophrenic Twin Studies (after Gottesman & Shields 1972)

Study	Numbers of pairs	% Schizophrenia & ? schizophrenia	% Sd 1 schizoid[a]	% Sd 2 other disorders[b]	% Normal
Luxenburger[c] (1928)	14	72	14	–	14
Rosanoff et al. (1934)	41	61	–	7	32
Kallmann (1946)	174	69	21	5	5
Slater (1953)	37	64	–	14	22
Kringlen (1967)	45	38	–	29	33
Fischer et al. (1969)	21	48	5	5	43
Gottesman & Shields (1972)	22	50	9	18	23

[a]So diagnosed by investigators.
[b]Includes as examples; alcoholic, psychopath (Kallmann); psychopathic; suicide (Slater), alcoholic, character neurosis (Kringlen).
[c]Only includes co-twins of certain schizophrenics.

185

Slater (1953) thought were distinctive of the abnormal relatives in schizophrenics' families. Employing a more liberal interpretation of the term schizophrenic-like, we considered from the data presented by Mosher et al. that at most 10 out of 16 of their co-twins could be so described. However, Mosher's twin sample was selected in a nationwide survey for MZ pairs where one member of the pair was an undoubted schizophrenic and the other was an undoubted nonschizophrenic. The selective biases thus introduced are hard to evaluate, but one result must certainly have been that the co-twins were an unusually healthy group, as in fact they are compared to the systematically ascertained twins.

Nevertheless, it cannot reliably be claimed, even when using very liberal criteria, either that nearly 100 percent of MZ co-twins are disordered (Sd 2) or that 100 percent are schizophrenic or schizoid personalities (Sd 1).

The Maudsley Schizophrenic Twin Series

It may be of interest to examine the recent Maudsley Hospital twin study from London (Gottesman & Shields 1972) with regard to possible pointers to a schizophrenic genotype (Sd 4) in the co-twins. We shall ask how many MZ and DZ co-twins of schizophrenics could be conservatively or liberally termed Sd 1 or Sd 2; and we shall discuss the findings in terms of 'schizoid disease' and polygenic theories. The results are shown in Table 5.

Of 22 MZ pairs in which a proband was definitely or probably schizophrenic according to the consensus of 6 diagnosticians, 11 (50 percent) were concordant for schizophrenia, using the same criteria for the co-twin. As shown below, 6 co-twins had nonschizophrenic consensus diagnoses. This gives an Sd 2 rate of 77 percent, using a conventional standard for 'disorder'. However, only 2 of these 6 co-twins had disorders resembling schizophrenia (Sd 1) or of a kind that would fall into the 'schizophrenia spectrum' of Kety et al. (1968).

Probably Schizoid Disorder (Sd 1)

   MZ 14B: male aged 20. Consensus diag: personality disorder (inadequate, hypochondriacal). Psychotic MMPI profile. Diagnosed pseudoneurotic schizophrenia by P. E. Meehl (P.E.M.) and as ? schizophrenia related personality by E. Essen-Möller (E.E-M.).

   MZ 16B: male aged 44. Consensus diag: personality disorder (paranoid). Refused MMPI. Schizotype (P.E.M.) and schizophrenia related personality (E.E-M.).

Probably Not Schizoid Disorder (Sd 1)

   MZ 5B: female aged 42. Consensus diag: neurotic depression, anxiety (suffered much stress; proband also anxious). Schizoidia never suspected.

Table 5. Disordered (Sd 2) and Possibly Schizoid (Sd 1) Co-twins of
Schizophrenics in the Maudsley Hospital Study

Classification of co-twins	MZ pairs	DZ pairs
a Schizophrenia or ?schizophrenia (consensus diag.)	11	3
b Other diagnosis: schizoid (clinical)	2	1
c Other diagnosis: ?schizoid (MMPI)	–	3
d Other diagnosis: no evidence of schizoidia	4	5
e Normal: ?Sd 4 schizoid (Essen-Möller)	3	–
f Normal: ?schizoid (MMPI and/or clinical)	–	3
g Normal: no evidence of schizoidia	2	18
Total	22	33
Sd 2, consensus a+b+c+d	17 (77%)	12 (36%)
Sd 1, maximum a+b+c+e+f	16 (72%)	10 (30%)
Sd 1 (maximum) or Sd 2 a+b+c+d+e+f	20 (91%)	15 (45%)

MZ 9B: female aged 47. Consensus diag: neurotic depression, anxiety.
Schizoidia never suspected, clinically or psychometrically.

MZ 18B: female aged 37. Consensus diag: personality disorder (hysterical), anxiety. Hospitalized 51 weeks for neurosis at age 20. P.E.M. diag.
pseudoneurotic schizophrenia, but E.E-M. did not relate personality to
schizophrenia, and schizoidia not otherwise suspected. Refused MMPI.
(Co-twin also anxious.)

MZ 24B: female aged 40. Consensus diag: post-partum depression. Hospitalized briefly at 31. Normal personality and MMPI.

The above may underestimate Sd 1. In addition to the 2 co-twins with
probably schizoid disorders, there were 3 psychiatrically normal co-twins
who had a personality which, according to Essen-Möller, may have been
indicative of a schizophrenic genotype (Sd 4), bringing the total of Sd 1
schizoids more broadly defined up to 5. The 11 schizophrenic and 5 possibly
schizoid Sd 1 co-twins together account for 72 percent of the 22 MZ pairs.
There was one co-twin who, at age 39, had an MMPI which might have led
to a suspicion of schizo-affective psychosis, if she had not otherwise appeared entirely unremarkable. The other normal MZ co-twin at age 46
shows nothing at all schizophrenic-like according to clinical impression,
diagnosticians' comments, or psychological tests; in this pair the proband
suffered from what might be regarded as a 'symptomatic' schizophrenia related to thyroid disease.

In the Maudsley study the premorbid personality of the first (schizophrenic) twin was, so far as the authors could judge, schizoid or probably so

in only 8 of 22 pairs. It is well known that many schizophrenias develop in personalities that are not Sd 1 schizoid (c.f. p. 178); we should not expect all MZ co-twins of schizophrenics to be schizoid in the sense of Sd 1.

Of the DZ pairs, 3 out of 33 (9 percent) were concordant for schizophrenia. Nine co-twins (3 of them hospitalized) had other disorders; these were mostly anxiety or depression, sometimes mild and transient. They bring the Sd 2 rate to 36 percent. Some of these other disorders, it was thought, could be accounted for by environmental stress or by depressive or other nonschizoid personality traits shared with the proband. Only one was generally regarded as a schizoid personality (DSM II sense), but 3 others among the 9 appeared to have a noteworthy schizoid element when the MMPI was considered. Among the normal co-twins a further 3 might be regarded as schizoid on MMPI and/or other evidence. The maximum concordance for schizophrenia or schizophrenic-like personality (Sd 1) is therefore 10/33 (30 percent) in DZ pairs.

The MZ and DZ concordance rates for Sd 2, 77 percent and 36 percent respectively, and those for Sd 1, 72 percent and 30 percent, fall short of the 100 percent and 50 percent required according to the dominant 'schizoid disease' hypothesis (Heston 1970), which makes an effort to avoid the introduction of incomplete manifestation. However, if we count co-twins who either are 'disordered' or are normal, but with a possibly schizoid personality, maximum rates of 20/22 (91 percent) MZ and 15/33 (45 percent) DZ are achieved, which fall very little short of those predicted. The critic may say, however, that the prediction of 50 percent affected DZ co-twins makes no allowance for the frequency of a gene supposed to be associated with the abnormalities that gave rise to the maximum rates. For example, if the posited schizoid gene had a frequency in the population of 10 percent, the expected risk in siblings (including DZ co-twins) of schizophrenics would be 56 percent rather than the 50 percent expected with rarer dominant gene conditions. The maximum rates of 91 percent in MZ and 45 percent in DZ pairs reported in Table 5 should also allow for the possibilities of environmental phenocopies and of genocopies causing some of the wide range of traits counted as "affected." In other words, allowance should be made for false positives contributing to the nice fit with dominant gene theory.

The alternative polygenic model—even less disprovable, according to its critics—looks for a well-defined measure which gives good discrimination between MZ and DZ concordance rates (Sd 3, MZ vs DZ), or which gives consistent estimates of heritability (not necessarily 100 percent!) independently derived from different kinds of relatives. In the context of the Maudsley study best agreement with the model was achieved with the diagnosis of schizophrenic psychosis itself, using standards that would be regarded as narrow in the U.S. and broad in the U.K. Shields and Gottesman (1972) have argued that similar diagnostic standards also work well in other recent twin studies. The search now should be for important contributory etiological factors; on the genetic side there may be no *single* Sd 4 indicator.

Schizoid Personality as Contributory Factor?

An hypothesis concerning Sd 1, in keeping with the observation that not all schizophrenics are premorbidly schizoid, is that a schizoid personality is just one of many factors, even if an important one, which can contribute in different proportions in different cases to the development of schizophrenia (Zerbin-Rüdin 1967). Once the genetic and environmental contributing factors result in the crossing of a threshold the schizophrenic psychosis ensues. Gottesman and Shields (1967) showed that the polygenic threshold model of Falconer (1965) worked reasonably well for schizophrenia. Developing ideas from Thoday's (1967) work on polygenic inheritance in animal genetics, they have subsequently (Gottesman & Shields 1972) suggested that factors from several different sources and with different weights contribute to the liability (Fig. 3).

How schizoid personality traits and schizoid psychopathy (Sd 1) will ultimately fit into this model remains to be seen. They could be either

Fig. 3. Schematic Representation of Various Genetic and Environmental Contributors to the Liability to Schizophrenia (Gottesman & Shields 1972).

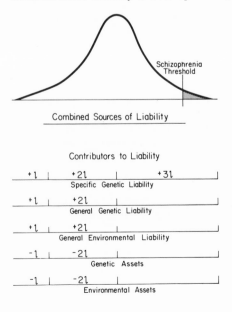

among the more specific or more general contributory *causes* with a strong genetic component, or they could be seen as subthreshold *effects* of other more fundamental causes—the result of the dynamic interplay of diathesis and stress in the relatively highly predisposed individual. They would be part of the zone of spectrum disorders shown in Figure 4.

There are a few pointers from other studies about the place of schizoid personality (Sd 1) in preschizophrenia. While there is quite good evidence from studies of normal twins (Shields 1973) that the personality dimension of introversion-extraversion has a significant genetic component, persons at the introversion extreme (? Sd 1) do not appear to be at high risk for developing schizophrenia. When shy, withdrawn children seen in child guidance clinics are followed up (Morris, Soroker, & Burriss 1954; Michael, Morris, & Soroker 1957) few are found to be schizophrenic. Child guidance cases who did become schizophrenic had more often presented with incorrigibility, running away from home, theft, and poor school performance (Robins 1966).

Gardner (1967) found obsessive-compulsive traits more often among child-guidance patients who became schizophrenic (particularly boys) than among patients who subsequently made a socially adequate adjustment. According to Watt, Stolorow, Lubensky, and McClelland (1970) unsocialized aggression was the most prominent pattern of behavior in preschizophrenic boys and (again with poorer discrimination from controls) over-inhibition in girls. Rutter's (1972) review considers that the premorbid abnormalities of schizophrenics do not constitute a sufficiently distinctive pattern for predictive purposes; and that there may be a dichotomy between schizophrenics who have difficulties in childhood and those who do not.

Fig. 4. Schematic Proposal for How Diathesis Interacts with Stress in the Ontogenesis of Schizophrenia (after Gottesman & Shields 1972).

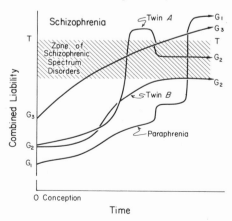

Of perhaps greater interest would be further studies based on schizoid psychopaths as probands. Earlier work, such as that of Stumpfl (1935) on cold, affectless psychopaths, did not reveal an increased familial incidence of schizophrenia; nor did recent work on sociopathic probands (Guze, Wolfgram, McKinney, & Cantwell 1967) or psychopaths from the Danish psychiatric register (Schulsinger 1972). Family studies have not been done on subjects identified in population surveys as high on the psychotic scales of the MMPI, but Hathaway, Monachesi, and Salasin (1970) found that only one of about 500 children more than 2 S.D. above the mean on the Schizophrenia or Paranoia scales had been hospitalized by the age of 25.

In Manfred Bleuler's (1972) long-term study of Burghölzli schizophrenics the sibs who became schizophrenic did not all do so on the basis of a schizoid disposition any more than the probands did. There was a clear tendency, however, for the schizophrenic sib to resemble the proband as to whether he was premorbidly a schizoid psychopath or not. The finding suggests that schizoid personality may be an independent contributory dimension with a genetic component rather than a manifestation of an essential unitary etiological factor.

The polygenic hypothesis, as does a dominant gene hypothesis with modifying genes, allows for individuals at genetic risk for developing a schizophrenic psychosis to vary in the degree of that risk. Some Sd 2 and Sd 3 disorders can be construed as less specific manifestations of "schizophrenic" genotypes, depending on the degree of genetic loading and relevant life experiences. Some discordant MZ pairs may be relatively low-risk subjects and one twin may have been more stressed than the other (c.f., Fig. 4). The nonspecific manifestations need not always represent high-risk schizophrenic genotypes.

Present and Earlier Views

Despite the different diagnostic and strategic procedures and the contradictory findings and interpretations to which we have drawn attention in our discussion of selected recent studies, perhaps the most promising pointers toward Sd 4 remain in the area in Figure 1 where our Sd 1, 2, and 3 circles overlap: it is the broadly schizophrenic-like disorders, such as schizoid character and borderline schizophrenia, which most consistently distinguish the relatives of process schizophrenics from appropriate controls. Beyond that little can be said. After stretching our resources to their utmost, we are no nearer to identifying other possible high-risk genotypes in schizophrenics' families. In the absence of good objective criteria for schizoid character and other 'borderline' conditions and the consequent lack of adequate epidemiological and family studies of such conditions, it cannot be claimed that we have an improved phenotype for population genetic studies.

Lessons from Diseases Better Understood Than Schizophrenia

At this point it might be helpful to see what can be learned from genetic diseases that are more completely known than schizophrenia. First, diabetes, which in its population genetics is remarkably like schizophrenia. As in schizophrenia, 45–50 percent of the MZ co-twins of affected persons are concordant. Then, if diabetes is defined as an abnormality in a glucose tolerance test (sometimes performed after an evocative stimulus), some of the remaining co-twins will be "chemically" diabetic (Gottlieb & Root 1968). Finally, if the plasma insulin response of co-twins who still seem normal is measured, it appears that nearly 100 percent will have a measurable abnormality (Cerasi & Luft 1967; Pyke, Cassar, Todd, & Taylor 1970). One problem here is that of genetic *expression*. What level of expression will we define as disease? What trait—overt diabetes, an abnormal glucose tolerance curve, or the plasma insulin response—is the best one for genetic analysis? Because severity of disease generally turns out to be important in medical genetics, perhaps all three traits will be useful, depending on one's purposes. And even plasma insulin levels are distant from gene action, so there will no doubt be other levels of trait definition to come. Although we think that the analogy to diabetes gives much to ponder that we will not make explicit, we will make the main points that expression of a genotype can vary widely indeed and that the comparatively extremely crude level at which the phenotype is assessed in schizophrenia is reason for humility and openness, not dogmatism.

A second useful example is the Lesch-Nyhan syndrome. This bizarre X-linked syndrome features a severe neuromuscular disorder, self-mutilation, and mental retardation. The defective enzyme (hypoxanthine guanine phosphoribosyltransferase) has about 0.005 percent of normal activity in the erythrocytes of affected persons. Now the activity of the same enzyme is deficient in an extraordinary range of other disorders found in Lesch-Nyhan families as well as families located through probands with one or another of those other disorders. Enzyme activity in the range of 0.01 percent to 0.5 percent of normal is associated with neurological disorders ranging from retardation to spino-cerebellar syndromes of variable severity. Levels of about 1 percent of normal are associated with gout (Kelley & Wyngaarden 1972; Seegmiller 1972). Certainly some of this clinical and biochemical variability will be associated with different mutations causing different amino acid substitutions in the same enzyme. Some variability can be attributed to differential modification of the enzymes' activity by environmental factors and by the balance of the genome. That the amount of protein translated from a mutant locus varies *between* families implicating modifying factors of the sort needed has been demonstrated in the case of sickle cell hemoglobin (Nance & Grove 1972). The main point is a simple one. There would be no possible way on clinical grounds to group all of the clinical disorders associated with deficiencies in the activity of this one enzyme into one clinical syndrome, not even those disorders appearing in one family. Again, tentativeness and humility are prescribed, but there is the further point that familial clustering of disease provides a logical classification, even if the diseases are very dissimilar at the phenotypic level.

Conclusion—Need for an Endophenotype

From what we have said about semantics, sampling, and other problems, we may not expect ready agreement about the most valid indicators of the schizoid state genetically (Sd 4). We should certainly strive for some better indication of 'what is inherited' than a mid-Atlantic diagnosis of classical schizophrenia. But without further advances in the basic biological sciences, the testing of promising leads will be a laborious and, some might think, a fruitless proceeding. It involves the lengthy follow-up of strategic populations relevant to the transmission of schizophrenia. We have mentioned the desirability of prospective investigation of loosely schizophrenic-like (Sd 1) and Sd 4-suspect subjects (e.g., the thought-disordered or eccentric or physiologically over-reactive) to discover how many of them and their relatives develop definite schizophrenic psychoses; and we need to know what becomes of the offspring of the matings of couples both of whom suffer from suspected schizophrenia spectrum disorders, both in known 'schizophrenic' families and in unselected samples from the general population.

Because doing such studies would expend prodigious labor to seek uncertain rewards, we think the best hope for the resolution of the schizophrenia problem may have to await the finding of a protein which differentiates schizophrenics from others. Short of that, close genetic linkage to a marker gene might be a possibility, despite the acknowledged difficulties (Jayakar 1970). However, the difficulty in distinguishing "affected" from "unaffected" family members, particularly in the younger age groups, would be likely to upset the arithmetic of linkage calculations (too many Sd 4's who are not even Sd 1 or Sd 2); the attempt might also founder because on a polygenic model there would be too many genes and on a monogenic model too many extraneous influences on expression.

Our hopes lie more with an endophenotype associated with the pathogenesis of schizophrenia. The characteristics of a good biological indicator of this kind have been outlined by Shields and Gottesman (1973). It should be stable enough to be reliably measured pre- and post-morbidly, as well as during a florid psychotic phase. A quantitative trait analogous to blood sugar level in diabetes would be useful, particularly if correlations between relatives suggested that it might be a highly heritable polygenic trait. Equally, if not more useful would be a polymorphism which segregated as a simple Mendelian trait. The rapid advances in the discovery of genetic variation in enzymes (Harris 1970) is a hopeful development. Depending on its relative frequency in schizophrenics, an enzyme variant might reveal itself as anything ranging from a minor polygenic contributor (like blood group 0 in duodenal ulcer) to a major specific factor in a polygenic system or (if rare and highly penetrant) the cause of a distinct class of schizophrenia entailing a high risk for relatives. If the Mendelian endophenotype were present in virtually all schizophrenics and in only some 3 percent of the general population, the monogene predicted by Slater's theory (Slater & Cowie 1971) would be found. No need *then* to worry about hard-to-define Sd 1, questionably circular Sd 2, or sampling-dependent Sd 3: we would have a direct Sd 4 indicator!

A biological advance may give us a better chance of solving some of the genetic problems in schizophrenia and the schizoid than juggling with clinical categories and test scores. A reliable biochemical measure might help to decide between competing genetic models, discover what psychopathology should be described as schizoid, and identify individuals at high risk of developing a malignant psychosis. In principle, a better understanding of genetic etiology should lead to improved environmental methods of treatment and prevention.

References

Alanen, Y. O. The family in the pathogenesis of schizophrenic and neurotic disorders. *Acta Psychiatrica Scandinavica*, 1966, Suppl. 189.

American Psychiatric Association. *Diagnostic and statistical manual of mental disorders.* (2nd ed.) Washington, D.C.: Committee on Nomenclature and Statistics of the American Psychiatric Association, 1968.

Bleuler, M. *Die schizophrenen Geistesstörungen im Lichte langjähriger Kranken- und Familiengeschichten.* Stuttgart: Thieme, 1972.

Brugger, C. Die erbbiologische Stellung der Pfropfschizophrenie. *Zeitschrift für die gesamte Neurologie und Psychiatrie*, 1928, **113**, 348–378.

Cerasi, E., & Luft, R. Insulin response to glucose infusion in diabetic and non-diabetic monozygotic twin pairs. Genetic control of insulin response? *Acta Endocrinologica*, 1967, **55**, 330–345.

Essen-Möller, E. Psychiatrische Untersuchungen an einer Serie von Zwillingen, *Acta Psychiatrica et Neurologica Scandinavica*, 1941, Suppl. 23.

Essen-Möller, E. The concept of schizoidia. *Monatsschrift für Psychiatrie und Neurologie*, 1946, **112**, 258–271.

Falconer, D. S. The inheritance of liability to certain diseases, estimated from the incidence among relatives. *Annals of Human Genetics*, 1965, **29**, 51–76.

Fischer, M., Harvald, B., & Hauge, M. A Danish twin study of schizophrenia. *British Journal of Psychiatry*, 1969, **115**, 981–990.

Gardner, G. G. The relationship between childhood neurotic symptomatology and later schizophrenia in males and females. *Journal of Nervous and Mental Disease*, 1967, **144**, 97–100.

General Register Office. A glossary of mental disorders. *Studies on Medical and Population Subjects*, No. 22. London: Her Majesty's Stationery Office, 1968.

Gottesman, I. I., & Shields, J. A polygenic theory of schizophrenia. *Proceedings of the National Academy of Sciences*, Washington, D.C., 1967, **58**, 199–205.

Gottesman, I. I., & Shields, J. *Schizophrenia and genetics: a twin study vantage point.* New York: Academic Press, 1972.

Gottlieb, M. S., & Root, H. F. Diabetes mellitus in twins. *Diabetes*, 1968, **17**, 693–704.

Guze, S. B., Wolfgram, E. D., McKinney, J. K., & Cantwell, D. P. Psychiatric illness in the families of convicted criminals: a study of 519 first-degree relatives. *Diseases of the Nervous System*, 1967, **28**, 651–659.

Hallgren, B., & Sjögren, T. A clinical and genetico-statistical study of schizophrenia and low-grade mental deficiency in a large Swedish rural population. *Acta Psychiatrica et Neurologica Scandinavica*, 1959, Suppl. 140.

Harris, H. *The principles of human biochemical genetics.* Amsterdam: North-Holland, 1970.

Hathaway, S. R., Monachesi, E., & Salasin, S. A follow-up study of MMPI high 8, schizoid, children. In M. Roff & D. Ricks (Eds.), *Life history research in psychopathology.* Minneapolis: University of Minnesota Press, 1970, pp. 171–188.

Heston, L. L. Psychiatric disorders in foster home reared children of schizophrenic mothers. *British Journal of Psychiatry*, 1966, **112**, 819–825.

Heston, L. L. The genetics of schizophrenic and schizoid disease. *Science*, 1970, **167**, 249–256.

Hutchings, B. Environmental and genetic factors in psychopathology and criminality. Unpublished M. Phil. Thesis, University of London, 1972.

Inouye, E. Similarity and dissimilarity of schizophrenia in twins. In *Proceedings of the Third International Congress of Psychiatry, 1961.* Vol. 1. Montreal: University of Toronto Press, 1963, pp. 524–530.

Jayakar, S. D. On the detection and estimation of linkage between a locus influencing a quantitative character and a marker locus. *Biometrics*, 1970, **26**, 451–464.

Jones, M. B. IQ and fertility in schizophrenia. *British Journal of Psychiatry*, 1973, **122**, 689–696.

Kallmann, F. J. *The genetics of schizophrenia.* New York: Augustin, 1938.

Kallmann, F. J. The genetic theory of schizophrenia: An analysis of 691 schizophrenic twin index families, *American Journal of Psychiatry*, 1946, **103**, 309–322.

Kelley, W. M., & Wyngaarden, J. B. The Lesch-Nyhan syndrome. In S. B. Stanbury, J. B. Wyngaarden & D. S. Fredrickson (Eds.), *The metabolic basis of inherited disease.* (3rd ed.) New York: McGraw-Hill, 1972, pp. 969–991.

Kety, S. S., Rosenthal, D., Wender, P. H., & Schulsinger, F. The types and prevalence of mental illness in the biological and adoptive families of adopted schizophrenics. In D. Rosenthal & S. S. Kety (Eds.), *The transmission of schizophrenia.* Oxford: Pergamon, 1968, pp. 345–362.

Kringlen, E. *Heredity and environment in the functional psychoses.* London: Heinemann, 1967.

Lindelius, R. (Ed.) A study of schizophrenia. *Acta Psychiatrica Scandinavica*, 1970, Suppl. 216.

Luxenburger, H.: Vorläufiger Bericht über psychiatrische Serienuntersuchungen an Zwillingen. *Zeitschrift für die gesamte Neurologie und Psychiatrie*, 1928, **116**, 297–326.

Meehl, P. E. Schizotaxia, schizotypy, schizophrenia. *American Psychologist*, 1962, **17**, 827–838.

Meehl, P. E. Manual for use with checklist of schizotypic signs. Minneapolis: University of Minnesota Medical School, 1964.

Meehl, P. E. A critical afterword, In I. I. Gottesman & J. Shields, *Schizophrenia and genetics — a twin study vantage point*. New York and London: Academic Press, 1972, pp. 367–415.

Meehl, P. E. MAXCOV-HITMAX: a taxonomic search method for loose genetic syndromes. In P. E. Meehl, *Psychodiagnosis: selected papers*. Minneapolis: University of Minnesota Press, 1973, pp. 200–224.

Michael, C. M., Morris, D. P., & Soroker, E. Follow-up studies of shy, withdrawn children. II: Relative incidence of schizophrenia. *American Journal of Orthopsychiatry*, 1957, **27**, 331–337.

Morris, D. P., Soroker, E., & Burruss, G. Follow-up studies of shy, withdrawn children. I. Evaluation of later adjustment. *American Journal of Orthopsychiatry*, 1954, **24**, 743–754.

Mosher, L. R., Pollin, W., & Stabenau, J. R. Families with identical twins discordant for schizophrenia: some relationships between identification, thinking styles, psychopathology and dominance-submissiveness. *British Journal of Psychiatry*, 1971, **118**, 29–42.

Mosher, L. R., Stabenau, J. R., & Pollin, W. Schizoidness in the non-schizophrenic identical cotwins of schizophrenics. In R. de la Fuente & M. N. Weisman (eds.), *Proceedings, V World Congress of Psychiatry, Mexico City, 28 November-4 December, 1971*. Amsterdam: Excerpta Medica, International Congress Series No. 274, 1973, pp. 1164–1176.

Nance, W. E., & Grove, J. Genetic determination of phenotypic variation in sickle cell trait. *Science*, 1972, **177**, 716–717.

Planansky, K. Phenotypic boundaries and genetic specificity in schizophrenia. In A. R. Kaplan (Ed.), *Genetic factors in "schizophrenia."* Springfield, Ill.: Thomas, 1972, pp. 141–172.

Pyke, D. A., Cassar, J., Todd, J., & Taylor, K. W. Glucose tolerance and serum insulin in identical twins of diabetics. *British Medical Journal*, 1970, **4**, 649–651.

Rado, S. Theory and therapy: the theory of schizotypal organization and its application to the treatment of decompensated schizotypal behavior. In S. Rado (Ed.), *Psychoanalysis of behavior*, Vol. II. New York: Grune & Stratton, 1962, pp. 127–140.

Robins, L. N. *Deviant children grown up*. Baltimore: Williams & Wilkins, 1966.

Rosanoff, A. J., Handy, L. M., Plesset, I. R., & Brush, S. The etiology of so-called schizophrenic psychoses with special reference to their occurrence in twins, *American Journal of Psychiatry*, 1934, **91**, 247–286.

Rosenthal, D. Two adoption studies of heredity in the schizophrenic disorders. In M. Bleuler & J. Angst (Eds.), *The origin of schizophrenia*, Bern: Huber, 1971, pp. 21–34.

Rosenthal, D. Three adoption studies of heredity in the schizophrenic disorders. *International Journal of Mental Health*, 1972, **1(1/2)**, 63–75.

Rosenthal, D., Wender, P. H., Kety, S. S., Schulsinger, F., Welner, J., & Østergaard, L. Schizophrenics' offspring reared in adoptive homes. In D. Rosenthal
& S. S. Kety (Eds.), *The transmission of schizophrenia*, Oxford: Pergamon,
1968, pp. 377-391.

Rosenthal, D., Wender, P. H., Kety, S. S., Welner, J., & Schulsinger, F. The
adopted-away offspring of schizophrenics. *American Journal of Psychiatry*,
1971, **128**, 307-311.

Rutter, M. L. Relationships between child and adult psychiatric disorders. *Acta
Psychiatrica Scandinavica*, 1972, **48**, 3-21.

Schulsinger, F. Psychopathy: heredity and environment. *International Journal of
Mental Health*, 1972, **1(1/2)**, 190-206.

Seegmiller, J. E. Lesch-Nyhan syndrome and the X-linked uric acidurias. *Hospital Practice*, April 1972, 79-80.

Shields, J. Concepts of heredity for schizophrenia. In M. Bleuler & J. Angst
(Eds.), *The origin of schizophrenia.* Bern: Huber, 1971, pp. 59-75.

Shields, J. Heredity and psychological abnormality. In H. J. Eysenck (Ed.),
Handbook of abnormal psychology (2nd ed.), London: Pitman Medical, 1973,
pp. 540-603.

Shields, J., & Gottesman, I. I. Cross-national diagnosis of schizophrenia in twins.
Archives of General Psychiatry, 1972, **27**, 725-730.

Shields, J., & Gottesman, I. I. Genetic studies of schizophrenia as signposts to
biochemistry. In L. L. Iversen and S. P. R. Rose (Eds.), *Biochemistry and
mental illness.* London: Biochemical Society, 1973, Special Publication 1, pp.
165-174.

Slater, E. (with the assistance of J. Shields). Psychotic and neurotic illnesses in
twins. *Medical Research Council Special Report Series* No. 278. London: Her
Majesty's Stationery Office, 1953.

Slater, E., & Cowie, V. A. *The genetics of mental disorders.* London: Oxford University Press, 1971.

Stumpfl, F. *Erbanlage und Verbrechen.* Berlin: Springer, 1935.

Thoday, J. M. New insights into continuous variation. In J. F. Crow & J. V. Neel
(Eds.), *Proceedings of the Third International Congress of Human Genetics.*
Baltimore: The Johns Hopkins Press, 1967, pp. 339-350.

Tienari, P. Psychiatric illnesses in identical twins. *Acta Psychiatrica Scandinavica*,
1963, Suppl. 171.

Watt, N. F., Stolorow, R. D., Lubensky, A. W., & McClelland, D. C. School adjustment and behavior of children hospitalized for schizophrenia as adults.
American Journal of Orthopsychiatry, 1970, **40**, 637-657.

Wender, P. H., Rosenthal, D., Kety, S. S., Schulsinger, F., & Welner, J. Social
class and psychopathology in adoptees: a natural experimental method for
separating the roles of genetic and experiential factors. *Archives of General
Psychiatry*, 1973, **28**, 318-325.

Wynne, L. C. Methodologic and conceptual issues in the study of schizophrenics
and their families. In D. Rosenthal & S. S. Kety (Eds.), *The transmission of
schizophrenia.* Oxford: Pergamon, 1968, pp. 185-199.

Zerbin-Rüdin, E. Endogene Psychosen. In P. E. Becker (Ed.), *Humangenetik,
ein kurzes Handbuch*, Vol. V/2. Stuttgart: Thieme, 1967, pp. 446-577.

13 DISCUSSION: THE CONCEPT OF SUBSCHIZOPHRENIC DISORDERS

David Rosenthal

When Dr. Kety, Dr. Wender, and I decided to explore and utilize the concept of a spectrum of schizophrenic disorders, we were not unaware that we might be opening another Pandora's Box, of which there are already so many in studies of mental illness. Nevertheless, there was historical precedent for what we planned to do, and the literature on the genetics of schizophrenia suggested that not all cases that might reflect a manifestation of a schizophrenia genotype were being counted.

Today's panel of papers indicates that the concept of subschizophrenic syndromes has attracted a lot of attention and stimulated much reexamination of this old problem. Not only has this issue aroused interest in the United States, but it has even become a major focus of concern in the Soviet Union. Professor A. V. Snezhnevsky (1972), who is head of the Institute of Psychiatry in Moscow, an Institute which is dedicated entirely to research on schizophrenia, is making research on the spectrum concept a central focus. He has coined the terms "nosos" and "pathos." He describes nosos as a morbid process, a dynamically developing formation, whereas pathos involves stable changes which are the result of the pathological process. Thus, the schizophrenic state is nosos, while some of the suggestive manifestations seen in the relatives of a schizophrenic, which have been described as the schizophrenic constitution, the schizoid, latent schizophrenia, and residual schizophrenia are called pathos. Thus, pathos contains the possibility of a pathological process (nosos), on one hand, and residual signs on the other. Dr. Snezhnevsky's Institute will carry out its research in accord with this conceptual framework.

Rather than discussing the individual chapters on schizophrenia at any great length, I will present some relevant data of our own with respect to the validity of the schizophrenia spectrum concept.

Dr. Bender's study is an exceptional one in that she maintains contact with childhood schizophrenics until they reach adulthood, and this research

David Rosenthal, Ph.D., Laboratory of Psychology, National Institute of Mental Health, Bethesda, Maryland.

strategy is an important achievement by itself. It is clear that Dr. Bender had plenty of opportunity to observe and evaluate the psychiatric status of the children, and hopefully their relatives as well. The confirmatory diagnoses of the children are her own, I believe, but diagnoses of the children's relatives probably were in good part based on institutional records or diagnoses, or on agency reports. True, the cases were not selected originally for research purposes, but Dr. Bender has exploited her material well. Apart from the striking differences regarding schizophrenia in the relatives of index and control groups, she found several disorders which seemed to be associated with schizophrenia, especially sociopathic personality and neurotic or inadequate personality. Later, I will show you why I think that sociopathic personality does not belong in the spectrum, despite the findings in her own study and in Heston's (1966) study of foster children.

Dr. Kidd tries to identify the mode of genetic transmission in schizophrenia and the selective advantage that might maintain a single major schizophrenia gene in the population. He concludes that "searching for a selective advantage among unaffected relatives of schizophrenics seems futile." He also reports that "no precise 'correct' answer is possible" with respect to identifying the mode of action, "because of the high heterogeneity in the values reported for the incidence of affected relatives of schizophrenics." I would add too that he had to make assumptions that are *not* valid, such as random mating and zero correlation in environmental factors among relatives, and he assumed additionally that there are only two alleles at the presumed locus, which could well be wrong. Thus, he had a lot going against him. But maybe the most important factor leading to his gloomy conclusion is the omission from his calculations of those kinds of disorders that all other panelists are focusing on today. It may be that some day we will know enough about these disorders that he can include them in his equations.

The paper by Shields, Heston, and Gottesman is a long and detailed one which attempts "to clarify some semantic and logical problems which have plagued the concept of the schizoid." The authors do achieve some clarifications and distinctions, but these do not appear to me to be critical distinctions of the type that will contribute significantly to a resolution of the problem. Our immediate and most important task with respect to the genetics of schizophrenia is to identify diagnostic groups or specific traits which may be genetically related to schizophrenia. Our goal is to find some way of proving the hypothesis regarding the spectrum disorders, one way or the other. Incidentally, the most informative review of this subject is one by Planansky (1972) who refers to this issue as the "boundaries of schizophrenia." Unfortunately, the article is not published in a journal but in a forty-three dollar book that may not find a wide audience, and that would be a pity.

The report by Dr. Kety gives me my first chance to be a discussant of a paper of which I am a coauthor. However, I do not plan to take advantage of this unique arrangement. Let me say simply that in this study and in our others, we are trying to decide on whether or not our schizophrenia spectrum disorders really are what we label them. In the Kety study, which we call the Family Study, the findings strongly favor the view that diagnostic categories such as uncertain schizophrenia and borderline schizophrenia are genetically related to classical process schizophrenia, but schizoid personality is not. Thus, we now talk about a hard spectrum and a soft spectrum which, when you think of it, may be taking us down the path of Snezhnevsky's nosos and pathos.

Undaunted, I want to show you why I think that the soft spectrum, which includes a number of syndromes that we call schizoid, is indeed genetically related to process schizophrenia. First, however, I must point out that most recent studies that have captured the scientific community's imagination, e.g., Heston's (1966), Mednick and Schulsinger's (1968), and our own, (Rosenthal, Wender, Kety, Schulsinger, Welner, & Østergaard 1968) began with a proband schizophrenic who was also a parent. No effort was made to determine the diagnostic status of the second parent. Such a research strategy may be useful, but from a genetic standpoint, it is equivalent to Gregor Mendel beginning his investigation with one selected type of sweet pea plant in the F_1 generation, crossing it with other plants whose characteristics he knew nothing about, and then trying to relate all the characteristics he finds in the F_2 generation to those of the F_1 parent whose special characteristics were known to him. Imagine the confusion and folly of this procedure. A proper genetic study must be based on who mates with whom. In fact, although genetics has been defined in various ways, the simplest and perhaps best definition of genetics is: the science of matings. This definition is especially appropriate for human genetics, since we do not have control of such mating and we tend to seek other research strategies to circumvent the problem of not having selected matings available to us in the ways or numbers we would like.

For these reasons, we decided to obtain psychiatric interviews of the second parent who had mated with the schizophrenic or manic-depressive index parent to produce the index child who was adopted away. We call this the Co-Parent Study, which is just one aspect of the Adoptees Study.

We began the Adoptees Study by identifying all individuals in greater Copenhagen who had been given up for adoption at an early age to nonrelatives during 1924 to 1947. We then identified those adoptees who had a biological parent with a schizophrenia spectrum or manic-depressive disorder. The adopted away offspring of these parents became our index cases. Controls were selected from the remaining adoptees in the total pool of approximately 5,500. The criteria for selecting controls included: same sex, age, social class rearing, and age at transfer to the adopting family as the index case to which he was matched. Also, neither biological parent of the control was listed in the national psychiatric register as having a mental illness.

Both index and control adoptees were brought to the Kommunehospital in Copenhagen for two days of examination. Dr. Joseph Welner conducted the psychiatric interviews, which lasted three to five hours, and in the great majority of cases, he did not know whether the adoptee he examined was an index or a control case. As part of his examination and findings, Dr. Welner made a brief diagnostic formulation of each subject.

The diagnoses of the index parents were based on summaries of their hospital records as prepared by a Danish psychiatrist, who then made his own diagnosis. Dr. Wender, Dr. Kety, and I made our own diagnoses independently, and subsequently Dr. Wender and I agreed to a consensus diagnosis for each parent.

Dr. Bjørn Jacobsen conducted the psychiatric interviews of the co-parents. The co-parent names were interspersed with the names of subjects who were part of two other studies we were carrying out at the same time, and so he could not be certain that any particular subject belonged to the co-parent study. Even if he could, he had no clear hypothesis about what the co-parent diagnoses should be, and he could not therefore be biased in this respect. After we received the report that accounted the interview in detail, Dr. Wender and I independently and then consensually made our own diagnoses of the co-parent, at the same time remaining blind as to the diagnosis of the index parent and the index child.

Our diagnostic coding system follows that of the American Psychiatric Association Diagnostic Manual II. However, we have added to it our own conception of the schizophrenia spectrum disorders. These include: questionable and definite acute schizophrenic episode; schizophrenic personalities such as undifferentiated inadequate and subparanoid, and schizoid personalities, all of which we refer to as the soft spectrum; and the hard spectrum which includes borderline and chronic schizophrenia, questionable or definite. We have also assigned an identifying number to each diagnosis to facilitate recording of diagnoses, and we have numbered the diagnoses so that, in the main, the higher the number the worse the diagnosis. The schizophrenia spectrum begins with 62, which is "questionable acute schizophrenic episode," and ends with 99, which is "chronic schizophrenia, hebephrenic." Numbers below 62 indicate that the diagnosis is not in the schizophrenia spectrum.

At this point in our work, we have not defined all the spectrum disorders as clearly or in as detailed a way as we would like, but we do seem to share a common conception of the respective disorders in the spectrum. We hope, of course, to provide the desired definitions in the future, and we have made a beginning in this respect. However, it should be pointed out that any degree of unclarity in our definitions of the spectrum disorders serves to work against us, in the sense that it leads to unreliability of diagnosis and increased disagreement among ourselves.

In the Co-Parent Study there were 79 co-parents. Some had died or were not available for study. However, we obtained an intensive psychiatric interview and/or additional information on 54 co-parents to whom we could assign a diagnosis, a participation rate of almost 70 percent, which is appreciable, considering the relatively advanced ages of the co-parents. Table 1 shows the diagnoses of the index parents and addresses itself to the question: With whom do schizophrenics mate?

It can be seen that when the index parent is female, the co-parent is diagnosed as having either a spectrum or psychopathic disorder in 56 percent of the matings, the two disorders represented equally in the male co-parents. When the index parent is male, however, almost half of the female co-parents (45 percent) have a spectrum diagnosis, and only one has a diagnosis of psychopathic disorder.

Thus, at least two salient points are brought out in Table 1 which are relevant to our discussion today.

1. If we define a mating between two persons, each of whom has a spectrum disorder, as an assortative mating, then it appears that such assortative matings are probably occurring at an appreciable rate.

2. A study such as Heston's (1966), which begins with a proband schizophrenic mother and finds an elevated rate of psychopathic disorders in the offspring, cannot conclude that the psychopathy is genetically associated with schizophrenia, unless he can show that the male co-parents were not psychopaths themselves.

Table 2 presents tentative evidence to the effect that the soft spectrum disorders are indeed genetically related to process schizophrenia, and it does this in the most direct way possible, utilizing a proper genetic study in which both F_1 parent phenotypical characteristics are identified and the F_2 phenotypical characteristics described qualitatively and quantitatively.

Table 1. Schizophrenia Spectrum and Psychopathic Personality Disorders in Co-Parents of Index Adoptees

Diagnosis of index parent	Co-parent male			Co-parent female		
	Number	Spectrum disorders	Psychopathic disorders	Number	Spectrum disorders	Psychopathic disorders
S_1*	15	5	5	9	4	0
S_2	3	0	1	1	1	0
S_3	2	0	1	2	1	0
$S_1/$ or $S_3/$	6	0	2	2	2	0
$D_1/$	1	1	0	1	0	0
S_4	2	1	0	1	0	1
M	3	2	0	6	2	0
Total	32	9	9	22	10	1

*S_1 = chronic schizophrenia; S_2 = acute schizophrenia; S_3 = borderline schizophrenia; $S_1/$ or $S_3/$ = possible or probable chronic or borderline schizophrenia; $D_1/$ = doubtful schizophrenia; S_4 = mixed symptoms of schizophrenia and manic-depressive illness; and M = manic-depressive illness.

In Table 2 all index parents have a consensus diagnosis of chronic schizophrenia. The diagnoses of the co-parents are represented by our numerical code. All numbers between 65, which represents "questionable schizophrenic or schizoid personality," and 79, which represents "paranoid personality," are in the soft spectrum. Numbers 80 and above represent the hard spectrum. Numbers 01 to 61 represent normality, organic, neurotic, affective, and personality disorders. Numbers 62 and 63 represent questionable and definite acute schizophrenic reaction, respectively. Frequently, the co-parent diagnosis involves two diagnoses separated by a slash mark. This was done because symptoms were often diverse and multiple, representing more than one diagnostic category, and we thought that both should be represented. The first diagnosis is the one we thought of as most salient, or primary, and the other was secondary. Anyone with a diagnosis of 65 or higher, whether primary or secondary, is included as an assortatively mating co-parent. Almost all assortative matings in Table 2 involve co-parents with a *soft* spectrum diagnosis.

On the right hand side of the table are Dr. Welner's most relevant diagnostic statements about the index adoptee. These diagnostic formulations can be classified as representing spectrum disorders or not, and the

Table 2. Diagnosis of Offspring of Schizophrenics According to Whether the Co-Parent Has a Schizophrenia Spectrum Disorder or Not

Mating pattern				Offspring	
#	♂ index	X	♀ co-parent	Sex	Diagnosis
1265	S_1[a]	X	78	♀	Psychoinfantile anxiety-neurotic, hysteriform woman.
3005	S_1	X	70/64	♂	Sz[b] borderline or perverse (homosexual?, transvestite). Could break down with Sz episode.
2243	S_1	X	65	♀	S is a very extroverted and selfassertive hysteric, childishly selfcentered but completely without prepsychotic or schizoid characteristics.
4021	S_1	X	65/40 + 48	♀	A beginning paranoid schizophrenic; unusually poor contact, vague, autistic-like life.
4643	S_1	X	65	♂	Pronouncedly antiaggressive; taciturn, shy.
1145	S_1	X	80	♂	Unusually introverted and shy; disturbed thinking.
1236	S_1	X	80/70	♀	Schizophreniform borderline? Micropsychotic-like episodes.
1617	S_1	X	67/78	♂	Mildly character neurotic. No schizoidy or psychosis. Salesman personality.

Table 2. continued.

	Mating pattern			Offspring	
#	♀ index	X	♂ co-parent	Sex	Diagnosis
5332	S$_1$	X	67/21	♂	Schizophreniform borderline case; primary process-like thinking.
26	S$_1$	X	62/65	♀	Schizophrenic borderline, or even more serious than that.
1034	S$_1$	X	51 + 13/77	♂	A normal person without any nervous symptoms worth mentioning. No prepsychotic characteristics ascertained.
3298	S$_1$	X	51 + 13	♀	Neurotic personality (anxiety and hysterical). No prepsychotic features.
3796	S$_1$	X	51/13	♀	Mentally defective. Personality cannot be described with exactness, but no suspicion of psychosis or prepsychotic characteristics.
2093	S$_1$	X	51/42	♀	Hysterical personality.
5295	S$_1$	X	51 mild	♀	Neurotic personality with depressive, anxious, and hysterical characteristics. No schizoid or prepsychotic phenomena.
100	S$_1$	X	13/42	♂	Though not schizoid or pre Sz, rather vulnerable. Orally dependent. Depressive reactions.
360	S$_1$	X	43/30	♂	An above average mentally healthy person.
5245	S$_1$	X	20	♀	Phobic-anxious neurosis, with antiaggressive character structure.
2154	S$_1$	X	01	♂	Healthy. A psychosomatic case without neurotic symptoms or structure.
2177	S$_1$	X	01	♂	Sz in a narrow sense. Sz thinking, primitive impulse breakthrough, pale contact.

	♂		♀		
#	index	X	co-parent		
521	S$_1$	X	43/48	♀	Affective disorder. Secretiveness and reserve. Vacillating psychopathic personality?
2127	S$_1$	X	39	♂	Complicated character neurosis.
4618	S$_1$	X	40/32	♂	Sensitive, self-insecure; vague schizoid tendencies.
2250	S$_1$	X	36	♀	Psychoinfantile personality with anxiety and hysterical symptoms. Thought process Sz-like, with regression. Acute Sz reaction? Oneroid state?

[a]S$_1$ = chronic schizophrenia

[b]Sz = schizophrenia

reader may make his own judgments about them. At this point, we raise the crucial question: Will we find more spectrum disorders among the offspring of assortative matings than of nonassortative matings, even when almost all the assortative matings involve a co-parent with a diagnosis in the soft spectrum? If so, such a finding would suggest that the genetic input from the soft spectrum co-parent adds to the genetic input from the chronic schizophrenic index parent to foster increased schizophrenic manifestation in the offspring. Two judges, Dr. Wender and I, independently sorted the offspring diagnoses as either "spectrum" or "not spectrum." With respect to four subjects, one of the judges could not make a decision with confidence and called the diagnosis "doubtful." The results are shown in Table 3.

It can be seen in Table 3 that the frequency of spectrum disorders in the offspring is about three to five times as frequent when the co-parent has a spectrum diagnosis than when he does not. Similarly, the offspring are about three times more likely not to have a spectrum diagnosis if the co-parent has no spectrum diagnosis. Unfortunately, the numbers are small, but a Fisher exact probability test indicates that the differences are significant at the 0.025 level, whether the doubtful cases are included or not. Dr. Kety and Dr. William G. Lawlor also sorted the offspring diagnoses independently, and they too obtained statistically significant differences.

Thus, the findings favor the view that the spectrum disorders are indeed genetically related to process schizophrenia. We have additional data on the remaining matings, but we are not ready to present them at the present time. In fact, we are planning to review Dr. Welner's interview summaries in detail and to make diagnostic assessments based not only on his diagnostic formulations, but upon his entire summary.

However, the immediate question before us today involves an apparent disagreement between the Family Study findings reported by Dr. Kety and the findings I have just reported in the Adoptees Study with regard to the soft spectrum. Why should this difference occur at all? Probably the best explanation can be found in the data shown in Table 4. The large majority of subjects in the Family Study reported by Dr. Kety are half-sibs of the probands. The index cases in the Adoptees Study are offspring of schizophrenics. In Table 4, risk rates for first-degree relatives of schizophrenics are shown on the left, and for second-degree relatives on the right. Note how the rates fall off, sometimes precipitously, as we move from first- to second-degree relatives. For sibs, a median rate among various studies is about 10.4 percent, but for half-sibs the rate is about one-third that. The risk rates for children tend to be higher than the rates for sibs. The rates most relevant to a comparison of the Family Study and the Adoptees Study are the rate for half-sibs and the rate for children, the rate for the latter being about four times larger. Thus, the degree of manifestation should be about four times greater in the Adoptees Study than in the Family Study. This difference probably applies to the soft spectrum as well, in which case the

Table 3. Offspring Diagnosis as Spectrum or Not Spectrum According to the Diagnosis of the Co-Parent When All Index Parents Are Chronic Schizophrenics

	Diagnosis of offspring	
Co-parent diagnosis	Spectrum	Not spectrum
Spectrum	5 (6)*	3 (4)
Not spectrum	1	10 (11)

*Figures in parentheses include subjects whose diagnosis was called "doubtful" by one judge but "spectrum" or "not spectrum" by the other. One subject was called "doubtful" by both judges, and cannot be shown in the table.

Fisher Exact Probability Test:
    P = 0.024 (Ss called "doubtful" by one judge excluded)
    P = 0.015 (Ss called "doubtful" by one judge included)

Table 4. Median-Risk Rates for Schizophrenia in First- and Second-Degree Relatives of Schizophrenics—According to E. Zerbin-Rüdin (1972)— Morbidity Risk (%)

First-degree relatives		Second-degree relatives	
Parents	6.3	Grandparents	1.6
Children	13.7	Grandchildren	3.5
Siblings	10.4	Half-Sibs	3.5

decreased manifestation in the half-sibs would be sufficient to make the identification of spectrum cases in the half-sibs appreciably more difficult, and many cases may be obscured or lost. The fact is that, even in the Family Study, one of the three judges did successfully discriminate between the index and control relatives with regard to the soft spectrum. If the spectrum disorders were not related to schizophrenia, but functioned like other personality disorders, the probability of one judge in three making such a discrimination would be miniscule. Thus, I believe that the concept of a schizophrenia spectrum of disorders is valid throughout its length and that we will have to reckon with it not only in assessing the genetics of schizophrenia, but in achieving a better understanding of general mental health problems as well.

References

Heston, L. L. Psychiatric disorders in foster home reared children of schizophrenic mothers. *British Journal of Psychiatry*, 1966, **112**(489), 819–825.

Mednick, S. A., & Schulsinger, F. Some premorbid characteristics related to break-down in children with schizophrenic mothers. In D. Rosenthal & S. S. Kety (Eds.), *The transmission of schizophrenia.* London: Pergamon, 1968, pp. 267–291.

Planansky, K. Phenotypic boundaries and genetic specificity in schizophrenia. In A. R. Kaplan (Ed.), Genetic factors in "schizophrenia." Springfield, Ill.: Thomas, 1972, pp. 141–172.

Rosenthal, D., Wender, P. H., Kety, S. S., Schulsinger, F., Welner, J., & Østergaard, L. Schizophrenics' offspring reared in adoptive homes. In D. Rosenthal & S. S. Kety (Eds.), *The transmission of schizophrenia.* London: Pergamon, 1968, pp. 377–391.

Snezhnevsky, A. V. Nosos et pathos schizophrenia. In A. V. Snezhnevsky (Ed.), *Schizophrenia: multidisciplinary investigations.* Moscow: Academy of Medical Science, Medicine, 1972, pp. 5–15.

Zerbin-Rüdin, E. Genetic research and the theory of schizophrenia. In L. Erlenmeyer-Kimling (Ed.), *International Journal of Mental Health.* New York: International Arts and Sciences Press, 1972, **1**, pp. 42–62.

14 DISCUSSION: PAPERS ON THE GENETICS OF SCHIZOPHRENIA

H. Warren Dunham

Of these four papers, only Professor Kidd's stands alone in that it attempts to develop a mathematical model that would encompass the various factors that are necessary for maintaining a certain frequency of schizophrenia in a population of risk. The other three papers in one way or another attempt to examine the nature of the genotype by statistical counts of certain social and psychological conditions that can be observed in the families of the phenotypes.

Thus, Professors Kety, Rosenthal, Wender, Schulsinger, and Jacobsen have analyzed further the data that formed the basis for their first report (Kety, Rosenthal, Wender, & Schulsinger 1968). I found that report, concerning the frequency of schizophrenia in biological parents as against adoptive parents, most impressive in its design. In their current report they have attempted to locate and interview all of the relatives in the families of their index and control groups, so I assume that their results will not rest entirely on data found in agency and hospital records. This new and extended step merely strengthens their original finding. Thus, for Table 3, which shows the total schizophrenic spectrum in all identified relatives for both index and control groups, I tested for a significant difference between the proportions for the six possible combinations. As their earlier findings demonstrated, and as might be expected, the differences are significant when the index biological relatives are compared with the index adopted, index biological with controls biological, and index biological with controls adopted, but are not significant in the other three comparisons, namely, index adoptive vs. control biological, index adoptive vs. control adoptive, and control biological vs. control adopted.

Further, I note with some surprise, especially in the light of the above findings, that the count of relatives with inadequate or schizoid personality is equally distributed between index and control cases for both biological and adoptive parents. I note also that for this diagnosis there appeared to be as much agreement between the three raters as for the other diagnos-

H. Warren Dunham, Ph.D., Professor of Sociology, Wayne State University School of Medicine, Departments of Community Medicine and Psychiatry, Detroit, Michigan.

tic categories. Such a finding, if supported by subsequent investigations, would throw into question the advisability of including this diagnosis in the schizophrenic spectrum for future studies of this type that attempt to establish a genetic basis for schizophrenia. Such a finding also produces a raised eyebrow over the paper of Drs. Shields, Heston, and Gottesman in which they attempt to examine the significance of the schizoid diagnosis in any analysis of the genetics of schizophrenia.

Dr. Bender's investigation, even with a less formidable design, provides a certain support to the findings of the Kety team. Her follow-up studies of schizophrenic children are well known and hence make a contribution to our knowledge concerning the natural history of schizophrenia. In the current paper she has examined the frequency of schizophrenia, other psychiatric diagnoses and sociopathy among the parents, siblings, and collaterals of fifty of her original childhood schizophrenics and compared them with fifty controls, none of whom had ever had a diagnosis of schizophrenia. The finding of schizophrenia in the fathers, mothers, siblings, and collaterals of the index cases, but no schizophrenia in the family members of the controls is a stubborn fact that cannot be dismissed.

The approximately equal numbers of fathers and mothers given a sociopathic diagnosis among index and control cases is to be noted, for it suggests that this diagnosis has little relevance in a genetic study of schizophrenia. The excess of sociopaths among the collaterals of schizophrenic subjects points to the desirability of looking at a breakdown by sex, for this diagnosis is more likely to be given to males than to females. In fact, if Dr. Bender's findings are accepted as valid, they suggest that diagnostic categories other than schizophrenia are of uncertain value in this type of genetic study, for they probably do little more than reflect the impoverished family background of these patients that were utilized in both schizophrenic and control groups.

In Professor Kidd's paper, as well as in other reviews of research findings and strategies on the relationship between genes and abnormal behavior, reference is generally made to propositions that are viewed as methodological obstacles to the establishment of conclusive findings concerning a genetic foundation for schizophrenia. The evidence which has so far been amassed and to which Professor Kidd and others working in this field give emphasis establishes without much doubt that a firm biological foundation for schizophrenia exists. After pointing up this fact, which Kallmann emphasized over thirty years ago, the research workers in this area always confront two propositions. First, they report an indecisiveness and uncertainty concerning the true nature of schizophrenia. Is this a disease entity or a quantitative variation in behavior that can be observed in certain persons? Such behavior could have a continuity from a mild withdrawal from human contact to a completely bizarre, thought disordered break with reality. With this recognition there is frequently a call for the development of some test or biochemical identifying agent that will enable the research worker to have a 100 percent diagnostic agreement in all the cases that he includes in his sample.

Parenthetically, perhaps some of you noted a recent account (*The Lancet* 1972) of a World Psychiatric Symposium on Schizophrenia in London last November which was attended by 400 psychiatrists. This account was entitled: "Status Quo in Schizophrenia," and this seems to be the reality in spite of the thousands of studies of this disease that have accumulated from all methodological perspectives and from all over the world. At this symposium, the report states, Dr. Joshua Bierer remarked that billions of dollars had been spent studying a disease that did not exist. The writer goes on to state that if it does not exist, something apparently does, for there are some 60,000 patients hospitalized in Britain and 180,000 hospitalized in the United States who apparently have something the matter with them. Thus, by experience we recognize that something that we call schizophrenia does exist, even though we don't know exactly what it is.

Second, research workers in this area continually emphasize the proposition that schizophrenia is a resultant of the interaction of genetic and environmental factors. The researchers in genetics, as would be expected, are much clearer concerning the genetic factors than the environmental ones. Thus, Professor Kidd notes in his paper what others have also noted, that the variation in concordance percentages in monozygotic twins points up the fact that environmental factors are quite significant and must be taken into account.

Therefore, it is not unexpected that Professor Kidd in his research has attempted to develop a model that will take account of both sets of factors in a population of risk. However, his attempts to construct a single major locus model with respect to the penetrance of three genotypes with 0, 1, and 2 schizophrenogenic alleles becomes highly questionable, for it is based upon findings concerning the rate of concordance of monozygotic and dizygotic twins. Such findings pose questions in themselves because of the variability of the rates in the several studies and, as Professor Kidd notes, makes it impossible to decide between the multifactorial and single major locus models for explaining the genotype. Professor Kidd develops his model which emphasizes only two alleles and is based upon four parameters; the allele frequency, q; the relative position of the heterozygote on the liability scale, h'; the environmental factor around each genetoype mean, e^2; and the position of the threshold that divided the population into the affected and unaffected, T. This model allows for the operation of both genetic and environmental factors in the production of a phenotype.

Professor Kidd's attempt to measure environmental penetrance by measuring the distribution of liability around the hypothetical mean, plus the emphasis given by practically all geneticists working in this area, that the environmental factor must be considered in the production of the schizophrenic phenotype, provides the opportunity for a sociologist to consider somewhat in depth the nature of the environment and the environmental influence in schizophrenia. The lip service that is given to the role of environmental factors in schizophrenia by geneticists often appears to voice a rather simple and naive conception as to what constitutes the environment.

One can only observe that if genetical study of behavior traits in humans is complex, any consideration of environmental factors presents an equal complexity, if not more so. Anyone who has worked in this field recognizes clearly that the concept of environment encompasses not only the physical and cultural surroundings confronting people which can be described and observed objectively but also the extent and intensity to which these entities, both physical and cultural, have penetrated the psyches of all persons in a given culture, and thus make for a variety of contrasting emphases in behavior responses. A person's response to various stimuli which intrude upon him is in good measure determined by the meanings of external events and cultural patterns which he has internalized. Psychologists have often noted from their studies that our values under certain conditions determine our perceptions. It should be recognized that a person does not respond automatically to these stimuli but has the capacity, because of these internalized meanings, to check his responses and to be selective with respect to the stimuli to which he wants to respond.

What I am suggesting is that within the course of the socialization process which each human neonate experiences there is a building up of a body of experience constituting the foundation which receives the selected stimuli and into which the new stimuli penetrate. The person then selectively responds to new stimuli from this background of experience, thus giving a new emphasis to them in the internalization of his response. This in turn is a new phase of his experience which becomes cumulative and which represents different substrata of the personality at the different periods in a person's life. With respect to certain stimuli that can be evaluated as harmful, one can clinically examine experience that is vulnerable to the receptivity of harmful stimuli against experience which is not vulnerable as it will reject the harmful stimuli.

This account of the internalization of stimuli from the surrounding physical and cultural environment does not intend to deny the role of the unconscious. Certain experiences will be forgotten and repressed for a variety of reasons and their resurrection may or may not be useful to a given personality. This account also is not intended to neglect the uniformity of meanings that a given culture can provide. It is only to say that each person in a society absorbs a sufficient amount of cultural symbols to make collective life possible, but that in certain personalities these experiences that are common to all can become so distorted as to constitute what John Dollard (1934) forty years ago viewed as a "private version of culture." This "private version of culture" might eventually be shown to constitute the environmental contribution to the schizophrenic disorder.

Perhaps, it is relevant to mention three groups of researchers that have attempted to examine the social milieu with respect to the development of schizophrenia. The research strategies are significant because of the inconclusive character of the findings and the difficulty of proving the hypothetical propositions that have emerged from the research. I am referring to the work of the Lidz group at Yale (R. Lidz & T. Lidz 1949) with their emphasis on role distortion and the lack of reciprocity of husband-wife roles in families; the Bateson group (Bateson, Jackson, Haley, & Weakland 1956) with their emphasis on the double-bind hypothesis; and the Wynne team (Wynne, Ryckoff, Day, & Hirsch 1958) in Washington which emphasizes the quality of interaction and the disjointed communication that goes on between persons in the family. It is interesting to note that each one of these research teams in one way or another attempts to get at the subjective meanings that are present in the persons as they carry on their daily round of life in the family.

Much is made by some research workers of the role of stress factors in precipitating various kinds of disturbed behavior. Stress is no doubt a significant element in the life of everybody, but one must note that the responses to similar stress situations vary markedly among persons, because each one encounters the stress factor from a different experiential background. Perhaps, this is another way of saying that some people in the course of their experience have learned to cope more effectively with stress than others.

Most attempts to examine the influence of environmental factors with respect to behavior of the self are handicapped by two methodological errors. First, they generally have a retrospective character. That is, they begin with a person who presents a behavior problem and then take a backward look at his development, his experience, and his interaction within the family milieu from which a factor or set of factors is isolated that supposedly explains his behavior.

The second methodological difficulty is in general the failure to conduct these studies of the background of problem personalities with a control group. Here the aim would be to examine the respective responses of problem and nonproblem personalities to similar cultural milieus and events.

Perhaps the great mystery of human existence is not that some persons display marked maladjustments with respect to traumatic cultural milieus and/or events, but rather that the great majority of persons survive

such experiences in one way or another without displaying maladjustments that will interfere significantly with their functioning.

Drs. Shields, Heston, and Gottesman in their paper are concerned with
unraveling an old problem with respect to schizophrenia. This problem had
appeared earlier in the literature in the guise of an environmental
problem; that is, the question was asked: To what extent may milder
forms of emotional disturbance, schizoid traits, and neuroses be fore-
runners of schizophrenia? The authors of this paper are concerned with
the same problem, but they place it in a genetic context by asking how to
categorize these types of cases which appear among relatives of persons
afflicted with schizophrenia. They attempt, through reference to various
studies in the literature, to examine the issue as to whether these conditions can be regarded as having any genetic significance with respect to
the development of the schizophrenic disorder.

Thus, they begin with a four-fold classification of the term schizoid as it
appears in the literature. Incidentally, this categorization of the term
does not in my judgment add anything to the task that they set for themselves, for they might just as easily have depended on the conventional
clinical types of psychiatric disorders and counted them as they appeared
among relatives of various samples of schizophrenics. The authors have
amassed a voluminous amount of data and information which they have
taken from the literature, but which for the most part has not served to
clarify or bring order to the task that they have set for themselves. In my
judgment, if they had depended on the data which they had presented in
Tables 5, 6, and 7 they could have made the point just as effectively and
could have ignored the other material.

Perhaps in summary it is worth noting that the authors of this paper are
hung up on the same concerns that I mentioned in regard to Professor
Kidd's paper, namely, the need for a more accurate diagnosis of schizophrenia and, second, the need to take account of the role of environmental
factors utilized to explain the lack of a complete 100 percent concordance
rate in MZ co-twins. Thus, they see the need for some biochemical
identifying agent for schizophrenia, but are cautious, of course, about
predicting its possibility. And, finally, the authors appear to be hung up on
quantitative-qualitative conflict with respect to schizophrenia.

In general, the paper is somewhat lengthy and I do not think the
authors have been selective enough in handling the literature in order to
deal with the more crucial types of data that might establish a case for
counting these abnormal conditions as they appear in the relatives of
schizophrenics. Again, I think this paper emphasizes our need for some
more careful pedigree studies of schizophrenics and nonschizophrenics,
where all of these conditions can be reliably established and counted.

Perhaps I should conclude my discussion of these papers by pointing out
what I view as the type of studies that are needed if we are ever to

unravel the causal complexity of the genetic and environmental factors in the production of certain phenotypes. First, as I indicated above, I think we need more careful clinical studies of human experience of both schizophrenics and normals coming from similar cultural milieus. Second, with respect to schizophrenia, I think we need more careful accounts of the natural history of the disease as occuring in persons from diverse cultural milieus.

References

Bateson, G., Jackson, D., Haley, J., & Weakland, J. Toward a theory of schizophrenia. *Behavioral Science*, October 1956, **1**, 251–264.

Dollard, J., The psychotic person seen culturally. *American Journal of Sociology*, 1934, **39**, 637–648.

Kety, S. S., Rosenthal, D., Wender, P. H., & Schulsinger, F. The types and prevalence of mental illness in the biological and adoptive families of adopted schizophrenics. In D. Rosenthal and S. S. Kety (Eds.), *The transmission of schizophrenia*. Oxford: Pergamon, 1968, pp. 345–362.

The Lancet, November 25, 1972, No. 7787, pp. 1133–34.

Lidz, R., & Lidz, T. The family environment of schizophrenic patients. *American Journal of Psychiatry*, February 1949, **106**, 332–345.

Wynne, L., Ryckoff, I., Day, J., & Hirsch, S. Pseudo-mutuality in the family relations of schizophrenics. *Psychiatry*, May 1958, **21**, 205–220.

IV

GENETIC STUDIES IN MANIC-DEPRESSIVE ILLNESS

15

LINKAGE STUDIES IN AFFECTIVE DISORDERS: THE Xg BLOOD GROUP AND MANIC-DEPRESSIVE ILLNESS

Julien Mendlewicz, Joseph L. Fleiss,
and Ronald R. Fieve

Introduction

Heredity has long been recognized as an important contributing factor in the etiology of manic-depressive illness (Slater 1936; Stenstedt 1952; Kallmann 1954). Specific studies have shown that monozygotic twins have a higher concordance rate for the disease than dizygotic twins (Luxenburger 1930; Slater 1953; Harvald & Hauge 1965; Kringlen 1967), that the relatives of manic-depressives are more likely to suffer from the disease than individuals in the population at large (Leonhard, Korff, & Schulz 1962; Perris 1966; Angst & Perris 1968; Winokur & Clayton 1967), and that the disorder often runs in two generations of a manic-depressive's family. Although these studies have established a genetic basis for manic-depressive psychosis, the mode of transmission is still unknown, and this has given rise to controversy centering around single-gene vs. polygenic inheritance. Furthermore, it is not yet clear whether the disease is homogeneous, or whether it encompasses different genetic subgroups.

Rosanoff, Handy, and Plesset (1935), in a twin study of manic-depressive illness, proposed that two dominant genes, one X-linked, were involved

Julien Mendlewicz, M.D., Ph.D., Department of Medical Genetics, New York State Psychiatric Institute; Research Associate in Psychiatry, Columbia University, New York, New York.

Joseph L. Fleiss, Ph.D., Biometrics Research Unit, New York State Department of Mental Hygiene; Associate Professor of Biostatistics, Colombia University, New York, New York.

Ronald R. Fieve, M.D., Chief of Psychiatric Research, Department of Internal Medicine, and Director, Lithium Clinic, New York State Psychiatric Institute; Associate Professor of Clinical Psychiatry, Columbia University, New York, New York.

The authors wish to acknowledge Dr. J. D. Rainer for constructive criticisms of the manuscript and M. Cataldo, M. A., for valuable research assistance. This investigation was supported in part by the Belgian American Educational Foundation and by General Research Support Grant 303-E165-F, New York State Psychiatric Institute.

in the transmission of the illness. More recently, Winokur and Tanna (1969) have also suggested that a locus for manic-depressive illness might be linked to the Xg blood group locus on the X chromosome. Reich, Clayton, and Winokur (1969) described one family which assorted for protan color blindness (an X-linked recessive genetic marker) and manic-depressive disease, and another which assorted for deutan color blindness (also X-linked) and manic-depressive illness. On the basis of close linkage in these two families, the authors hypothesized the existence of an X-linked factor in the transmission of manic-depressive illness. We have also reported significant linkage between the illness and both color blindness and the Xg blood group (Mendlewicz, Fieve, & Rainer 1971; Mendlewicz, Fleiss, & Fieve 1972b; Fieve, Mendlewicz, & Fleiss 1973).

In the present communication, we report on 12 additional families assorting for the Xg blood group and manic-depressive illness. The probands were all from a sample of over 185 carefully diagnosed manic-depressive patients attending the Lithium Treatment and Research Clinic at the New York State Psychiatric Institute. Probands and their relatives were studied from January 1971 to September 1972.

Subjects and Methods

The sample at the Lithium Treatment and Research Clinic consists of about 50 unipolar patients in addition to the over 135 bipolar patients. All patients as well as their spouses and their available first-, second-, and third-degree relatives on both the maternal and paternal sides were personally examined. When interviewing a proband's relative, the examiner was kept blind with respect to whether the proband was unipolar or bipolar.

The diagnoses of manic-depressive illness (bipolar depression) and depressive illness (unipolar depression) in probands were made separately by two investigators, following criteria similar to those of Leonhard et al. (1962) and Winokur, Clayton, and Reich (1969, p. 124). The Current and Past Psychopathology Scales (Endicott & Spitzer 1972), a semistructured clinical interview, was used for the evaluation of psychopathology in probands and relatives.

Bipolar depression was diagnosed in probands and relatives who had a history both of clear-cut manic behavior, and of depressive episodes severe enough to require treatment or hospitalization, or to cause a disruption in everyday activities for at least three weeks. Among the criteria used for this diagnosis was periodicity of illness with symptom-free intervals. Unipolar depression was diagnosed in individuals aged twenty or over who had never experienced mania or hypomania but had experienced one or more depressive episodes severe enough to require treatment or hospitalization. The lower age limit was taken at twenty to avoid confounding depression with childhood and adolescent disorders. For either bipolar or unipolar illness there had to be a lack of personality disintegration before and following psychotic episodes and no other preexisting psychiatric or medical disease which might be associated with an affective symptomatology.

The family study method, consisting of personal interviews with the relatives, was adopted because this method has been proved to be more reliable than the family history method, in which family history data is collected from the proband (Rimmer & Chambers 1969). In addition, pertinent medical and social records for probands and relatives were used when available. The material also reflects information based on recollections of subjects who were interviewed about unavailable relatives. This information has been used in family K (I(1)).

Five cc's of blood from each available subject were collected on E.D.T.A. and refrigerated before being sent in random order to the New York Blood Center for blood group analysis.* These analyses were performed blind to the subject's clinical condition.

Manic-depressive illness, like any phenotypic entity, may be influenced in its clinical manifestations by environmental factors. Thus, *within given families* sharing the genotype for manic-depressive illness, a predisposed individual can express either the unipolar or bipolar phenotype, depending on his internal and external environment. Even though a number of studies have indicated that unipolar and bipolar affective illness may be genetically distinct (Leonhard et al. 1962; Perris 1966; Angst & Perris 1968; Winokur & Clayton 1967), most of them (Leonhard et al. 1962; Angst & Perris 1968; Winokur & Clayton 1967) have found a higher than expected prevalence of unipolar illness in the relatives of bipolar probands. Furthermore, Zerbin-Rüdin (1969), in a review of twin studies in the literature, has shown that when both members of a monozygotic pair have an affective illness, in a high proportion of pairs one suffers from unipolar illness and the other from bipolar illness.

These findings might be explained by the effect of environment or personality on the outward expression of the bipolar genotype, thus suggesting that unipolar relatives of bipolar probands might have the bipolar genotype. Our assumption is, therefore, that *within a family unit* identified by a bipolar proband, bipolar illness and unipolar illness are genetically related and express the same genotype.

Twelve probands were ascertained whose pedigrees were informative for the analysis of linkage between bipolar illness and the Xg blood group. A pedigree was identified as informative whenever there was a demonstrably doubly heterozygous mother with two or more sons. Daugh-

---

*Blood group analyses were kindly provided by Dr. F. H. Allen of the New York Blood Center.

ters are informative when their father has the recessive phenotype at both loci. By doubly heterozygous we mean that the mother has both the dominant and recessive alleles for the illness and the blood group. In the majority of cases, double heterozygosity in the mother was established by some of her sons being well and others ill, and by some being Xg positive and others Xg negative. In other instances, heterozygosity in the mother could be established by her parents' phenotypes.

Linkage between pairs of traits can be assessed by estimating the frequency of recombination of these traits. During the process of meiosis in the mother, some material from one X chromosome may cross over and be exchanged for material from the other, i.e., recombination may occur (Fig. 1). If a female subject carries on one of her X-chromosomes the dominant alleles A, B, C, and if she carries on the other X chromosome the recessive alleles a, b, c, the relative frequency of recombination is a function of the distance between two loci. The closer the loci, like A and B, the more unlikely is recombination to occur. The more distant they are, like A and C, or B and C, the more likely is recombination to occur. Offspring whose phenotypes indicate that recombination occurred are called *recombinants*. Those whose phenotypes indicate no recombination are called *nonrecombinants*.

Results

In the first five pedigrees to be presented (P, MO, L, M, and K), the number of nonrecombinants and recombinants could be determined.

Family P—The proband's paternal aunt, II (5), has bipolar illness and is Xg positive. Individual II (5) is heterozygous for the illness, because two of her children, III (3) and III (5), are affectively ill, whereas, the other two are well, and she is heterozygous for the blood group because one of her children, III (4), is Xg negative and the others are Xg positive. This portion of the pedigree represents an instance where daughters are informative for X-linkage (usually, only sons are informative because they receive their single X chromosome from their mothers). The father, II (6), is well and Xg negative so that an ill daughter can only have received the gene for the

Figure 1

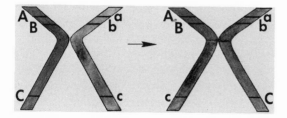

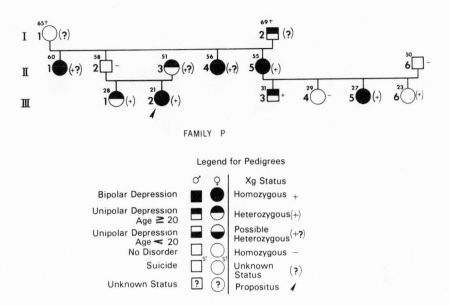

FAMILY P

Legend for Pedigrees

	σ	♀	Xg Status
Bipolar Depression	■	●	Homozygous +
Unipolar Depression Age ≥ 20	▣	◓	Heterozygous (+)
Unipolar Depression Age < 20	◩	◒	Possible Heterozygous (+?)
No Disorder	□ₛ₁	○	Homozygous −
Suicide	□ₛ₁	○ₛ₁	Unknown Status (?)
Unknown Status	?	?	Propositus ⟋

Figure 2

illness from her mother and an Xg-positive daughter can also only have received the Xg-positive gene from her mother.

The mother's brother, II (2), is well and Xg negative. This suggests that the mother, II (5), has the traits in *coupling*, i.e., that the alleles for the illness and Xg positive are on one X chromosome and that the alleles for no illness and Xg negative are on the other. Three of the offspring, III (3) to III (5), may, therefore, be counted as nonrecombinants because they are either ill and Xg positive or well and Xg negative. The fourth offspring, III (6), is counted as a recombinant because she is well but Xg positive.

Family MO—The proband, III (3), has bipolar illness and is Xg negative. He has one brother, III (1), who is Xg negative and also has the illness, and one sister, III (2), who had no evidence of the illness and is Xg positive. This represents another instance where a daughter is informative for X-linkage. The mother, II (2), is doubly heterozygous. Her brother, II (3), is Xg negative and has the illness, suggesting that the mother has the traits in *repulsion*, i.e., that the alleles for the illness and Xg negative are on one X chromosome and that the alleles for no illness and Xg positive are on the other. The three offspring III (1) to III (3) are therefore counted as three nonrecombinants because they are either ill and Xg negative or well and Xg positive.

Family L—The proband, III (6), is a female who is an identical twin. She has bipolar illness and is heterozygous for the Xg blood group. Her two sons are too young to be counted for purposes of statistical analysis. Her

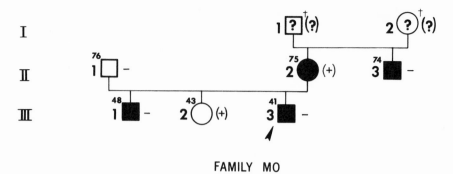

FAMILY MO

Figure 3

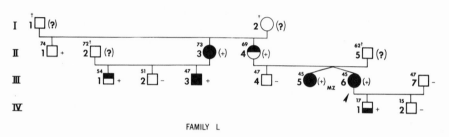

FAMILY L

Figure 4

maternal aunt, II (3), is bipolar and is heterozygous for Xg. This maternal aunt has three sons. III (1) is depressed and Xg positive. III (2) has never had an affective episode and is Xg negative; and III (3) is bipolar and Xg positive. Their mother's brother, II (1), is well and Xg positive, suggesting that the traits may be in repulsion. We therefore count III (1) to III (3) as three recombinants.

Family M—The proband has two sons, III (1) and III (2). One of them, III (1), is Xg positive and has been hospitalized for depression. The other son, III (2), is Xg negative and has no history of affective disorder. The characteristics of the proband's brother, II (1), suggest that the traits are in coupling. Therefore, the proband's sons are counted as two nonrecombinants.

Family K—The proband, III (2), has been hospitalized for manic-depressive illness and is Xg positive. His brother, III (3), also has manic-depressive illness but is Xg negative. Their mother, II (2), has suffered from depression and is heterozygous for the illness because her father, I (1), has never been reported to be ill. She is obviously heterozygous for Xg.

Because the maternal uncle, II (1), refused to provide a blood test and because the maternal grandfather is dead, the linkage *phase* in the

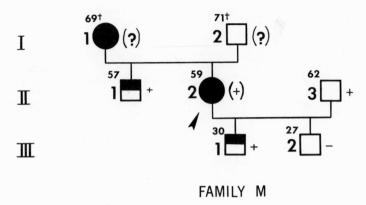

FAMILY M

Figure 5

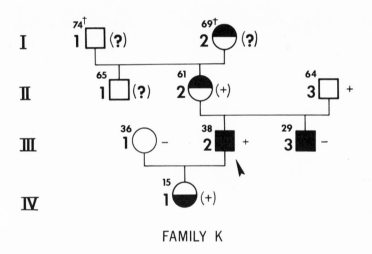

FAMILY K

Figure 6

mother (i.e., coupling or repulsion) cannot be determined. One of III (2) and III (3) is a nonrecombinant and the other a recombinant, although it is impossible to determine which is which.

In the next seven pedigrees, linkage phase could not be established and therefore the numbers of nonrecombinants and recombinants could not be determined. Nevertheless, these pedigrees did provide data for the mathematical analysis of linkage.

Family SA—The proband, II (3), is bipolar and Xg positive. His older brother, II (1), and his fraternal twin, II (2), have never had an affective illness and are Xg negative.

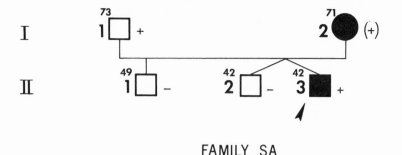

FAMILY SA

Figure 7

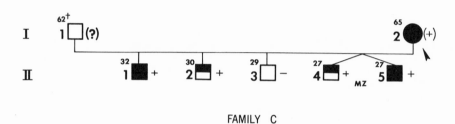

FAMILY C

Figure 8

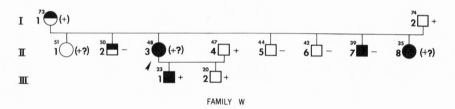

FAMILY W

Figure 9

Family C—The proband, I (2), has five sons. Two of them, II (4) and II (5), are monozygotic twins. They are therefore genetically identical and cannot be counted as two separate individuals, but rather as one. Individuals II (1), II (2), and the pair of twins are affectively ill and Xg positive; individual II (3) is well and Xg negative.

Family W—The proband, II (3), has four brothers. One of them, II (2), is unipolar and Xg negative. Two of her other brothers, II (5) and II (6), are well and Xg negative. Her remaining brother, II (7), is bipolar but also Xg negative.

Families R, S, F, CO—Each of these families provides two informative sons, one ill and Xg positive, the other well and Xg negative.

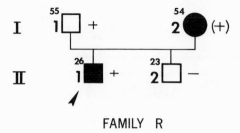

FAMILY R

Figure 10

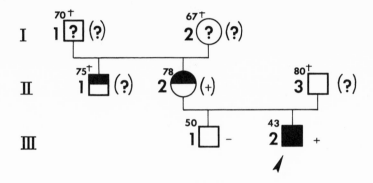

FAMILY S

Figure 11

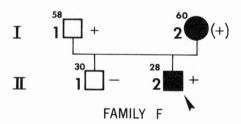

FAMILY F

Figure 12

Analysis

The degree of linkage between a pair of traits is estimated by the recombination fraction, θ. The minimum value of θ is zero, which indicates that the loci for the two traits coincide. The maximum value is 0.50, which indicates either that the loci for the two traits are on opposite ends of the same chromosome or that they are on two different chromosomes. The closer the value of θ is to zero, the stronger is the evidence for linkage.

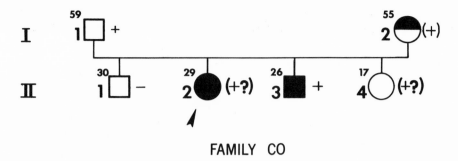

FAMILY CO

Figure 13

The simplest estimate of θ is the ratio of the number of recombinants to the total number of informative individuals, recombinants plus non-recombinants. An improved estimate is provided by the maximum likelihood method of Morton (1955) as tabulated by Edwards (1971). Morton's method is superior because it incorporates the means used to ascertain double heterozygosity and linkage phase in the mother. The method gives greater weight to the data from the four families in which phase could be determined (P, MO, L, and M) than to the data from the remaining eight families.

Table 1 presents, for each of our twelve pedigrees and the two informative pedigrees of Winokur and Tanna (1969), the logarithm of the odds favoring the indicated value of θ over the value 0.50. Logarithms are taken to facilitate the combination of evidence from many pedigrees.

The maximum likelihood estimate of θ is $\hat{\theta} = 0.19$, with a standard error of 0.07. The upper 95 percent confidence limit for θ is 0.30, meaning that we can assert with 95 percent confidence that the recombination fraction is no greater than 0.30. The odds favoring a recombination fraction of $\theta = 0.19$ over the value 0.50 are appreciable, nearly 70 to 1.

The evidence is therefore good for measurable linkage between manic-depressive illness and the Xg blood group. The estimated value of the recombination fraction is significantly less than 0.50, and the odds in its favor are over 60 to 1.

A possible source of difficulty in genetic studies of manic-depressive illness is the differential diagnosis between unipolar depression and schizo-phrenia in subjects under the age of forty. Many American psychiatrists would probably have diagnosed some of our young unipolars as schizo-phrenic (Cooper, Kendell, Gurland, Sharpe, Copeland, & Simon 1972). In our previous study (Mendlewicz et al. 1972b), we attempted to control for the possible diagnostic confusion between unipolar illness and schizo-phrenia by considering as definite unipolars only those over the age of

Table 1. Analysis of Linkage between Manic-depressive Illness and the Xg Blood Group

Family	Method of ascertainment	a+d*	b+c*	Recombination fraction θ								
				.05	.10	.15	.20	.25	.30	.35	.40	.45
P	CMBC	3	1	-.18	.05	.14	.18	.17	.15	.10	.06	.02
MO	CMBR	0	3	.81	.74	.66	.57	.47	.36	.24	.13	.04
L	CMBR	3	0	-1.47	-.92	-.62	-.42	-.28	-.18	-.10	-.05	-.01
M	CMBC	2	0	.52	.47	.42	.36	.29	.22	.14	.07	.02
K	CS2	1	1	-.76	-.47	-.32	-.21	-.14	-.08	-.05	-.02	0
SA	CS2	3	0	.51	.45	.38	.31	.24	.17	.10	.05	.01
C	CS2	4	0	.80	.71	.62	.51	.41	.30	.19	.09	.03
W	CS2	2	2	-1.45	-.90	-.59	-.39	-.25	-.15	-.08	-.04	-.01
R	CS2	2	0	.22	.19	.15	.12	.09	.06	.03	.02	0
S	CS2	2	0	.22	.19	.15	.12	.09	.06	.03	.02	0
F	CS2	2	0	.22	.19	.15	.12	.09	.06	.03	.02	0
CO	CS2	2	0	.22	.19	.15	.12	.09	.06	.03	.02	0
W6†	CS2	3	0	.51	.45	.38	.31	.24	.17	.10	.05	.01
W7†	CS2	4	1	-.19	.02	.10	.12	.12	.09	.06	.03	.01
Sum of log odds				-.02	1.36	1.77	1.82	1.63	1.29	.82	.45	.12
Antilog = odds favoring θ over .50				1.0	23	59	66	43	19	6.6	2.8	1.3

*a = number of offspring ill and Xg positive
b = number of offspring ill and Xg negative
c = number of offspring well and Xg positive
d = number of offspring well and Xg negative
†Winokur & Tanna 1969.

forty. Applying the same method of control to the current data, we estimate the recombination fraction to be 0.29, with a standard error of 0.08 (odds = 2.7 to 1). Measurable linkage is thus weaker, but is still statistically significant.

This kind of control is overly strict and unrealistic, however. Almost every family study of manic-depressive illness reported the morbidity risk for unipolar illness in relatives to be high, but the risk for schizophrenia to be low (Kallmann 1954; Angst & Perris 1968; Winokur et al. 1969; Slater 1936). It is for this reason that we have discontinued separating unipolars into those under forty and those over forty.

Discussion and Conclusions

Fourteen families informative for linkage between manic-depressive illness and the Xg blood group have been studied, two reported by Winokur and Tanna (1969) and twelve in this report. Strong evidence for measurable linkage was found. This analysis lends further support to the previous linkage studies of Winokur and Tanna (1969) and Reich et al. (1969), and the more recent linkage reports from our laboratory by Mendlewicz et al. (1971), Mendlewicz et al. (1972b), and Feive et al. (1973), all indicating dominant X-linked transmission in the families studied.

The Xg positive allele, in contrast to other X-linked markers (color blindness, G-6-PD deficiency, etc.), occurs in close to half of all males in the general population (65 percent). Therefore, other investigators can, with a high likelihood, ascertain informative families assorting for the Xg marker and the illness in order to test the hypothesis of X-linkage in manic-depressive illness.

Despite our findings consistent with X-linked dominant transmission in manic-depressive illness, there may still be genetic heterogeneity (Mendlewicz, Fieve, Rainer, & Fleiss 1972a; Mendlewicz, Fieve, Rainer, & Cataldo 1973; Mendlewicz & Rainer 1973). That is, it is still possible that in some subgroups multiple alleles at one locus or an involvement of different alleles at different loci are needed to explain the genetic characterization of this illness. In addition, there may be polygenic inheritance in other subgroups (Slater, Maxwell, & Price, 1971; Perris, 1971). None of these models can be ruled out, and one cannot generalize from the data presented that all cases of manic-depressive illness are X-linked. We may, however, conclude that within the families presented here, a dominant X-linked gene is involved in the transmission of manic-depressive illness.

References

Angst, J., & Perris, C. Zur Nosologie endogener Depression: Vergleich der Ergebnisse zweier Untersuchungen. *Archiv für Psychiatrie und Nervenkrankheiten,* 1968, **210**, 373–386.

Cooper, J. E., Kendell, R. E., Gurland, B. J., Sharpe, L., Copeland, J. R. M., & Simon, R. *Psychiatric diagnosis in New York and London.* London: Oxford University Press, 1972.

Edwards, J. H. The analysis of X-linkage. *Annals of Human Genetics*, 1971, **341**, 229–250.

Endicott, J., & Spitzer, R. L. Current and past psychopathology scales: rationale, reliability, and validity. *Archives of General Psychiatry*, 1972, **27**, 678–687.

Fieve, R. R., Mendlewicz, J., & Fleiss, J. L. Manic-depressive illness: linkage with the Xg blood group. *American Journal of Psychiatry,* 1973, **130**, 1355–1359.

Harvald, B., & Hauge, M. Heredity factors elucidated by twin studies. In J. V. Neel, M. W. Shaw, & W. J. Schull (Eds.), *Genetics and the epidemiology of chronic diseases.* Public Health Service Publication 1163. U.S. Department of Health, Education and Welfare, Washington, D.C. 1965.

Kallmann, F. J. Genetic principles in manic-depressive psychoses. In P. Hoch & J. Zubin (Eds.), *Depression.* New York: Grune & Stratton, 1954, pp. 1–24.

Kringlen, E. *Heredity and environment in the functional psychoses.* An epidemiological–clinical twin study. Oslo: Universitetsforlaget, 1967.

Leonhard, K., Korff, I., & Schulz, H. Die Temperamente in den Familien der monopolaren and bipolaren phasischen Psychosen. *Psychiatria et Neurologia,* 1962, **143**, 416–434.

Luxenburger, H. Psychiatrisch-neurologische Zwillingspathologie. *Zentralblatt für die gesamte Neurologie und Psychiatrie,* 1930, **56**, 145–181.

Mendlewicz, J., Fieve, R. R., & Rainer, J. D. Linkage studies in affective disorders. Paper presented at the Fifth World Congress of Psychiatry, Mexico City, December, 1971.

Mendlewicz, J., Fieve, R. R., Rainer, J. D., & Cataldo, M. Affective disorder on paternal and maternal sides. Observations in bipolar (manic-depressive) patients with and without a family history. *British Journal of Psychiatry*, 1973, **122**, 31–34.

Mendlewicz, J., Fieve, R. R., Rainer, J. D., & Fleiss, J. L. Manic-depressive illness: a comparative study of patients with and without a family history. *British Journal of Psychiatry*, 1972, **120**, 523–530, (a)

Mendlewicz, J., Fleiss, J. L., & Fieve, R. R. Evidence for X-linkage in the transmission of manic-depressive illness. *Journal of the American Medical Association*, 1972, **222**, 1624–1627. (b)

Mendlewicz, J., & Rainer, J. D. Transmission of manic-depressive illness. *Journal of the American Medical Association*, 1973, **224**, 1187.

Morton, N. E. Sequential tests for the detection of linkage. *American Journal of Genetics*, 1955, **7**, 277–318.

Perris, C. A study of bipolar (manic-depressive) and unipolar recurrent psychoses (I–X). *Acta Psychiatrica Scandinavica*, 1966, Supplement 194.

Perris, C. Abnormality on paternal and maternal sides: observations in bipolar (manic-depressive) and unipolar (depressive) psychoses. *British Journal of Psychiatry*, 1971, **118**, 207–210.

Reich, T., Clayton, P. J., & Winokur, G. Family history studies: V. the genetics of mania. *American Journal of Psychiatry*, 1969, **125**, 1358–1369.

Rimmer, J., & Chambers, D. S. Alcoholism: methodological considerations in the study of family illness. *American Journal of Orthopsychiatry*, 1969, **39**, 760–768.

Rosanoff, A. J., Handy, L. M., & Plesset, I. R. The etiology of manic-depressive syndromes with special reference to their occurrence in twins. *American Journal of Psychiatry*, 1935, **91**, 725–762.

Slater, E. The inheritance of manic-depressive insanity. *Proceedings of the Royal Society of Medicine*, 1936, **29**, 981–990.

Slater, E. Psychotic and neurotic illnesses in twins. *Medical Research Council Special Report Series, No. 278.* Her Majesty's Stationery Office, London, 1953.

Slater, E., Maxwell, J., & Price, J. S. Distribution of ancestral secondary cases in bipolar affective disorders. *British Journal of Psychiatry*, 1971, **118**, 215–218.

Stenstedt, A. A study in manic-depressive psychosis. *Acta Psychiatrica Neurologica Scandinavica*, 1952, Supplement 79, pp. 1–111.

Winokur, G., & Clayton, P. J. Family history studies: 1. two types of affective disorders separated according to genetic and clinical factors. In J. Wortis (Ed.), *Recent advances in biological psychiatry* (Vol. 9), New York: Plenum Press, 1967, p. 935.

Winokur, G., Clayton, P. J., & Reich, T. *Manic-depressive illness.* St. Louis: C. V. Mosby, 1969.

Winokur, G., & Tanna, V. L. Possible role of X-linked dominant factor in manic-depressive disease. *Diseases of the Nervous System*, 1969, **30**, 89.

Zerbin-Rüdin, E. Zur Genetik der depressiven Erkrankungen. In H. Hippius & H. Selbach (Eds.), *Das depressive Syndrom.* Munich: Urban & Schwarzenberg, 1969.

16 THE FAMILIAL HISTORY IN SIXTEEN MALES WITH BIPOLAR MANIC-DEPRESSIVE ILLNESS

H. Von Grieff, Paul R. McHugh,
and Peter E. Stokes

Introduction

The great interest in the genetics of manic depression in recent years has given rise to different views concerning the possible mechanism of transmission involved in this illness. Recently the possibility of an X-linked transmission has been again proposed. The reports of Perris (1968) and Angst and Perris (1968) claim X-linked dominant transmission for the unipolar variety but not for the bipolar. This view is opposed to that of Winokur, Clayton and Reich (1969) and Reich, Clayton and Winokur (1969) who propose an X-linked dominant transmission in the bipolar type. A study of male bipolar patients and their families is needed to evaluate these proposals.

Index Patients

The index patients were a series of 16 male patients admitted consecutively to the Lithium Clinic of The New York Hospital. All the index patients have a history of at least one manic attack, and at least on one occasion have had an attack of depression. Their histories were documented by interview as well as from the records of previous hospitaliza-

H. Von Greiff, M.D., Instructor of Psychiatry, Cornell University Medical College; Unit Supervisor, The New York Hospital-Cornell Medical Center, Westchester Division.

Paul R. McHugh, M.D., Professor of Psychiatry, Cornell University Medical College; Clinical Director and Supervisor of Psychiatric Education, The New York Hospital-Cornell Medical Center, Westchester Division.

Peter E. Stokes, M.D., Associate Professor of Psychiatry and Medicine, Cornell University Medical College; Chief of Psychobiology, Study Unit, Payne Whitney Clinic.

This work was supported in part by NIMH Grant #MH 12464 and the George Baker Trust. Also by the NIMH Graduate Residency Training Program in Psychiatry, Grant #2T01-MH11000-06. We would also like to acknowledge the enthusiastic support and help of Helen Goodell in collecting information for this paper.

tions. Thus, they were all found to be examples of bipolar manic-depressive disorder by the strictest criteria.

The age of onset of the disorder in these patients ranged from eighteen to thirty-nine, with a median age of thirty-four. These patients were thus all examples of early onset bipolar manic depressive disorder.

All these patients were treated with lithium and in the course of the last eighteen months only one patient has been rehospitalized with a reappearance of mania.

Relatives of Index Cases

Examples of affective disorder were sought among the first-degree relatives of these index cases. It was decided that the criteria for illness in a relative should be that the person: (1) was a suicide, or (2) received psychiatric treatment for a marked affective change as a hospitalized or office patient, or (3) because of conspicuous changes in mood his social adaptations had been impaired for two months or more. Emotional instability, brief mood swings, or transient depressions were not considered criteria for secondary cases.

Although information about the majority of ascertained cases was derived from the proband alone, in one-third of the ascertained cases support was sought from hospital records and direct interviews, which confirmed the impression derived from the proband. An attempt was made to find cases among second-degree relatives, but it was likely that our collection of these relatives was incomplete. However, the same techniques for questioning the proband were used to determine the presence of illness in the second-degree relatives, and again the results were confirmed by hospital records, corroborating relatives, or interviews in one-third of the cases. Occasionally, affected third-degree relatives were discovered. The third-degree relatives were recognized mostly if they suffered from a psychiatric disorder.

Results and Interpretation

Table 1 shows the recognized morbidity (uncorrected) of affective disorder in the relatives of our 16 male bipolar index cases. These 16 index cases have 90 first-degree relatives, of whom 20 were, by our criteria, examples themselves of affective disorder, yielding a frequency for affected first-degree relatives of some 22 percent. This is somewhat larger than the 12 percent found by Stenstedt (1952), but quite comparable to that reported by Hopkinson and Ley (1969) in patients with an onset of affective disorder before age forty, as was true of all of our index patients.

As revealed in the table, these first-degree relatives included 32 parents, of whom 8 were ill (incidence 25 percent), 34 siblings, of whom 9

Table 1. Recognized Morbidity (Uncorrected) of Affective Disorder in the
Relatives of Sixteen Male Manic Index Patients

Class of relatives	Number of relatives	Number with affective disorder
First-degree	90	20 (22%)
Parents	32	8 (25%)
Siblings	34	9 (26%)
Children	24	3 (12.5%)
Second-degree	111	9 (8.1%)
Grandparents	40	4
Half sibs	2	2
Uncles and aunts	30	2
Nephews and nieces	36	1
Grandchildren	3	0

were ill (incidence 26 percent), and 24 children, of whom 3 were ill
(incidence 12.5 percent). The index patients themselves are still so young
that few of their children have passed through the period of risk. We have
elected, though, not to correct for this feature for the purpose of this
communication.

Of the 111 second-degree relatives 9 had affective disorder for an
incidence of 8.1 percent.

These incidence percentages, as other workers have reported for patients with manic-depressive disorder, do not reveal a clear Mendelian
pattern of inheritance. However, we found, as have so many others, the
similar incidence in parents and siblings that excludes autosomal recessive inheritance as the sole mode of transmission of this disorder. The
percentages also exclude a pure dominant inheritance. However, they
could be used to indicate the presence of an autosomal gene expressing
itself in some fraction of the heterozygotes. This would be the view of
Kallmann (1954) and Stenstedt (1952). On the other hand, a polygenic
inheritance would seem equally plausible.

Of our 16 index patients we were able to search the parental lines back
beyond the father and mother for 9 of them. In these 9 families there was
no incidence of the disorder in either the maternal or paternal line in 3.
In 2 families there was incidence of the disorder in both the maternal and
paternal line. In 4 families the incidence was unilateral, restricted to the
paternal line in 3 and to the maternal line in 1. We realize that these are
not sufficient numbers to draw any clear conclusions as to transmission,
but the recognition that unilateral transmission is found in but a minority
of our cases suggests again the possibility that the condition is polygeni-
cally inherited.

The sex of the affected first-degree relative is of particular interest in this study, since we are choosing to look at male bipolar patients to test the issue of X-linked transmission. Of the 20 ascertained cases among the 90 first-degree relatives there were 14 males and 6 females. Of particular interest is the observation that of the 8 parental cases, 7 were fathers and 1 was a mother, and of the 3 affected children, 2 were sons and 1 a daughter. Looking at these affected relatives still more closely, we find that 3 of the 7 fathers were in fact hospitalized for bipolar affective disorder, 2 of them in our hospital. Four fathers were treated for depression only. The one mother has been hospitalized for bipolar illness. Of the 2 sons, one has been treated for mania, the other for depression. Now if the condition is to be inherited as an X-linked dominant, the affected parents and affected children of these males should be all female. Thus these results contradict the hypothesis that X-linked dominant inheritance is the sole mode of transmission of bipolar manic-depressive disorder.

It is of interest to look at some of the genealogies of these male patients who we believe take their genetic endowment from the paternal side. Case one (the index case) shows a typical son of a father who himself was hospitalized at our hospital for bipolar disorder, has been treated with electroconvulsive therapy, and has had his family disrupted many times through attacks of mania. The mother was totally free of any disorder, as were her parents. Several siblings of the index case also have shown manic-depressive disorder. The two older brothers committed suicide and the two sisters have been hospitalized with bipolar manic-depressive disorder.

Case two is extremely interesting because it confirms a paternal line of transmission of bipolar manic-depressive disorder. Our index case is the son of a father who has suffered from attacks of depression and other periods of irritability and impulsivity for which he has needed psychiatric

Fig. 1. Genealogy of Case 1, a Male with Bipolar Illness.

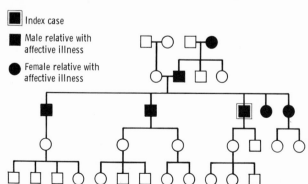

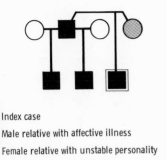

■ Index case

■ Male relative with affective illness

⊘ Female relative with unstable personality

Fig. 2. Genealogy of Case 2, a Male with Bipolar Illness.

treatment. This father has been married three times. With each of these three wives he has sired a son who has developed bipolar manic-depressive disorder. We have interviewed the proband and one of his half-brothers, confirming these features.

Thus from our families we can exclude the possibility of autosomal recessive transmission, or X-linked dominant transmission, or strict dominant transmission as being the only modes of inheritance of manic-depressive disorder of the bipolar type. It is conceivable that the condition is heterogeneously inherited. Manic-depressive disorder, as now recognized in bipolar form, could have a heterogenous transmission, some families inheriting in a dominant fashion, other families in a recessive fashion, still others as an X-linked condition. Some of the genealogies could be construed to show autosomal dominant, autosomal recessive, or X-linked dominant transmission. However, our data could most economically be viewed as the results of a polygenic inheritance.

Case 1: P.R., age 61, had his first hospitalization for depression in 1948 at age 39. He felt worthless, hopeless, and because of increasing suicidal ruminations he was given electroconvulsive treatments with improvement. He was hospitalized again in 1967 for depression with recurrence of suicidal ideas and psychomotor retardation. He again was given shock treatment with improvement. He also has had several periods of increased energy, sexual unrest, inability to sleep, overspending, but never required hospitalization for these attacks.

Family History: The patient's father died in 1937 in a psychiatric hospital. He had a history of episides of mania and depression with several hospitalizations in our hospital and each episode required shock treatment. Mother—no symptoms.

The patient's older sibling, a male, committed suicide at age 60 by shooting himself. Although he was never hospitalized, he was prone to moderate mood swings during which his behavior became noticeably different than his usual self.

Second sibling, a male, also committed suicide. He jumped out of a window at the age of 42. He is also described as having had mood swings, being elated and depressed, but was never hospitalized. A younger sister, age 59, had her first depression in 1933, which required hospitalization for 6 months. After her discharge she was noticed to have seasonal variations, feeling depressed in the fall and elated in the spring. These cycles persisted until the summer of 1936 when she became very active with increased energy and her mood was elated; however, during the fall of that year she became severely depressed and attempted suicide with barbiturates. This required hospitalization at the Westchester Division. Shortly thereafter, she became manic, being extremely active, overtalkative, and writing numerous letters. This manic state lasted a year. The patient was then transferred to another psychiatric institution where she went through several depressions and elations, trying several times to commit suicide while being depressed. This patient is presently alive, functioning as an out-patient, but, according to the proband, she is somewhat "wild."

The youngest sister was first hospitalized in 1963 for depression at the Westchester Division. She was given electroconvulsive treatments with improvement. After discharge she has had several episodes of elation, during which she exhibited increased energy, sexual drive, and an inability to sleep. She is presently hospitalized and has been so for the past two years at a psychiatric hospital for depression.

Paternal grandmother suffered a depression that lasted several months during the onset of her menopause.

Case 2: H.P., age 29, was first hospitalized at age 24 at our hospital for feelings of depression. He felt extremely suspicious with feelings of persecution that increased in intensity. His second hospitalization was in 1971 for a manic episode. The patient felt extremely confident, claimed to be able to accomplish anything he set himself to do. He would get extremely angry and violent if his wishes were thwarted.

Family History: The patient's father is age 91. He has suffered from attacks of depression and from periods of extreme irritability, impulsivity, and sexual unrest, for which he has sought psychiatric treatment. Lately he has been depressed for the past 3 months. He has been married three times. He has had 3 sons—one by each marriage. All 3 sons suffer from manic-depressive illness, bipolar type. The patient's mother is described by the patient and by the interviewer as a histrionic, anxious, demanding person, prone to temper outbursts and moodiness.

One of the two half-brothers, both older than the patient, wrote to us and described his periods of elation and depression for which he required hospitalization. He has been on Lithium for the past 3 years and is doing well. The other half-sibling we could not reach, but his 2 brothers and stepmother report him suffering from incapacitating attacks of depression and excitement, for which he has received psychiatric treatment.

Summary

Sixteen males suffering from bipolar manic-depressive illness consecutively admitted to the Lithium Clinic of the New York Hospital, and their families, were examined to test the issue of X-linked transmission. Twenty-two percent of the first-degree relatives of the index cases were found to be suffering from an affective disorder. No clear mendelian pattern of inheritance was found. There was evidence of father-to-son transmission of affective illness in four cases without incidence in the maternal line. These findings contradict the hypothesis of X-linked dominant inheritance as the sole mode of transmission of bipolar manic-depressive illness. The possibility of heterogenous transmission or of polygenic inheritance is suggested by our data.

References

Angst, J., & Perris, C. Zur Nosologie endogener Depressionen, Vergleich der Ergebnisse zweier Untersuchungen. *Archiv für Psychiatrie und Nervenkrankheiten*, 1968, **210**, 373–386.

Hopkinson, G., & Ley, P. A genetic study of affective disorder. *British Journal of Psychiatry*, 1969, **115**, 917–922.

Kallmann, F. Genetic principles in manic depressive psychosis. In P. H. Hoch & J. Zubin (Eds.), *Depression*. New York: Grune & Stratton, 1954.

Perris, C. Genetic transmission of depressive psychoses. *Acta Psychiatria Scandinavica*, 1968, Supplement 203, 45–52.

Reich, T., Clayton, P. J., & Winokur, G. Family history studies: V. the genetics of mania. *American Journal of Psychiatry*, 1969, **125**, 1358–1369.

Stenstedt, A. A study of manic depressive psychosis, clinical social and genetic investigations. *Acta Psychiatria Scandinavica*, 1952, Supplement 79.

Winokur, G., Clayton, P. J., & Reich, T. *Manic depressive illness*. St. Louis: C. V. Mosby, 1969.

A DOMINANT X-LINKED FACTOR IN MANIC-DEPRESSIVE ILLNESS: STUDIES WITH COLOR BLINDNESS

Ronald R. Fieve, Julien Mendlewicz,
John D. Rainer, and Joseph L. Fleiss

Clinicians have noted for decades that episodes of mania and depression tend to run in two or more generations of the same family (Kraepelin 1921). Thus heredity has long been implicated in the etiology of manic-depressive illness. To strengthen this clinical observation, computations of morbid risk (Slater 1936; Stenstedt 1952; Kallmann 1954) have shown that the relatives of manic-depressive probands are more likely to suffer from this illness than are members of the general population. In addition, twin studies (Kallmann 1954; Slater 1951; Zerbin-Rüdin 1969; Rosanoff, Handy, & Plesset 1935) have shown a greater concordance in monozygotic than dizygotic twin pairs, where one of each pair has the manic-depressive illness. These findings all point to a hereditary factor. A further and strongly compelling argument for a genetic factor in manic-depressive disease would be the establishment of linkage between the illness and one or more known genetic markers.

Linkage means the physical presence of the loci for two genes in close proximity on the same chromosome, one for a known gene, and the second for a gene either established or being investigated, as in the case of

Ronald R. Fieve, M.D., Chief of Psychiatric Research, Department of Internal Medicine, and Director, Lithium Clinic, New York State Psychiatric Institute; Associate Professor of Clinical Psychiatry, Columbia University, New York, New York.

Julien Mendlewicz, M.D., Ph.D., Department of Medical Genetics, New York State Psychiatric Institute; Research Associate in Psychiatry, Columbia University, New York, New York.

John D. Rainer, M.D., Chief of Psychiatric Research, Department of Medical Genetics, New York State Psychiatric Institute; Professor of Clinical Psychiatry, Columbia University, New York, New York.

Joseph L. Fleiss, Ph.D., Biometrics Research Unit, New York State Department of Mental Hygiene; Associate Professor of Biostatistics, Columbia University, New York, New York.

This study was supported by the Lithium Clinic Grant # C58173 from the Division of Local Services, Department of Mental Hygiene, Albany, New York.

manic-depressive illness. In the field of human genetics a number of linkage studies have demonstrated by means of statistical techniques that a given trait or illness is linked to a known genetic marker. In the case of the X chromosome, the sex distribution in pedigrees itself can suggest X-linked inheritance. To date it has been possible to identify over 86 traits carried on the X chromosome as reported in McKusick's (1971) latest catalogue. A few of these are frequent enough to be used as markers for the more precise measurement of linkage, and about 10 pairs of loci have been found to be within measurable distance of each other.

The first linkage studies undertaken in manic-depressive illness were those of Tanna and Winokur (1968), who obtained no evidence of linkage with the ABO blood system (autosomal in nature) in a study of both unipolar and bipolar families. In a second investigation, Winokur and Tanna (1969) did find evidence for linkage in three families assorting for manic-depressive illness and the Xg blood group, an X-linked trait.

Simultaneously, Reich, Clayton, and Winokur (1969) reported on two manic-depressive families in which color blindness, also an X-linked trait, seemed to be linked with bipolar illness. In the first of a series of linkage studies undertaken in our laboratory, Mendlewicz, Fleiss, and Fieve (1972b) reported on seven families assorting for manic-depressive illness and either deutan or protan color blindness. Statistically significant linkage was found. This report reviews and extends our previous study of color blindness, which demonstrated X-linkage in some manic-depressive families. In addition it presents new evidence for genetic heterogeneity in that a limited number of pedigrees have also been detected in which there appears to be father-son transmission.

Subjects and Methods

The sample was obtained between January 1971 and October 1972 by screening for color blindness all 50 unipolar and all 120 bipolar affective disorder patients attending the Lithium Treatment and Research Clinic at the New York State Psychiatric Institute. From this sample 11 families were ascertained as segregating for both color blindness and bipolar manic-depressive illness. Of these 11 families, 9 were informative. All probands, as well as their spouses and available first-, second-, and third-degree relatives were examined blindly for the presence of color blindness and psychopathology. The diagnosis of bipolar manic-depressive illness in probands was made independently by two investigators using Winokur's criteria (Winokur, Clayton, & Reich 1969). Bipolar illness (manic-depression) was diagnosed when there were clearcut symptoms of manic and depressive behavior.

Criteria used for the diagnosis of mania (Winokur et al. 1969) included a euphoric mood with at least four of the following symptoms: grandiosity, extravagance, over-talkativeness, flight of ideas, increased motor activity, and decreased need for sleep. Criteria required for the diagnosis of depression (Winokur et al. 1969) included a rapid or gradual onset of feelings of depression with at least three of the following symptoms: loss of energy; loss of interest in friends, social activities, or work; sleep disturbance; anorexia; loss of libido; motor retardation; decreased concentration; and suicidal ideas. An episodic nature to the illness, symptom-free intervals with relatively normal interval functioning, and no chronic deterioration were required for diagnosing both bipolar and unipolar illnesses. A clinical semistructured interview (Endicott & Spitzer 1972) was also used for confirming the diagnosis and for evaluation of the current and past psychopathology in probands and relatives.

As we have noted in previous reports, some studies have indicated that unipolar and bipolar illnesses may be genetically different (Leonhard, Korff, & Shulz 1962; Winokur & Clayton 1967; Angst & Perris 1968; Perris 1966) although most of these (Leonhard et al. 1962; Winokur & Clayton 1967; Angst & Perris 1968) have found a higher than expected prevalence of unipolar illness in the relatives of bipolar probands. This fact, along with the review by Zerbin-Rüdin (1969), which showed that in a high proportion of monozygotic twin pairs with affective illness one member suffered from bipolar illness and the other from unipolar illness, led us to assume in this report that within a family pedigree identified by a bipolar proband, bipolar and unipolar illness are genetically related and express the same genotype.

A standard illumination source was used to test for color vision. All subjects were shown all plates and every female was tested independently for each eye to discover any possible mosaicism. Two methods were used for testing color-vision anomalies: the Hardy–Rand–Rittler test, 2nd ed. (Hardy, Rand, & Rittler 1957), a true pseudoisochromatic test; and the Farnsworth–Munsell-100-Hue Test (Farnsworth 1947), based on the ability to match hues that progress around the color circle in 100 steps. These tests of color vision were always performed after the psychiatric evaluations in order to reduce the possibility of bias entering into the psychological assessments.

Description of Kindreds

Protan Color Blindness (Figure 1)
Family M—The proband, II (2), has two sons. One of them, III (1), is color blind and has been hospitalized for depression. The proband's other son, III (2), is neither color blind nor depressed. The fact that the mother's brother II (1), is ill and color blind suggests that the two traits are in coupling. We therefore consider that III (1) and III (2) are two nonrecombinants.

Family N—The proband's maternal grandfather is ill and color blind. The traits are therefore in coupling and the proband can be counted as a nonrecombinant.

Family S—The proband's maternal uncle, II (1), has been depressed and color blind, suggesting that the traits are in coupling. The proband, III (2), and his brother, III (1), may therefore be counted as two non-recombinants.

Family P—The proband has a first cousin, III (1), who is color blind and has bipolar depression. Because the maternal grandfather was ill and color blind the phase is in coupling. III (1) can therefore be counted as a nonrecombinant.

Fig. 1. Pedigree with Protan Color Blindness. (Reprinted with permission from Mendlewicz, Fleiss, & Fieve 1972*b*, p. 1626.)

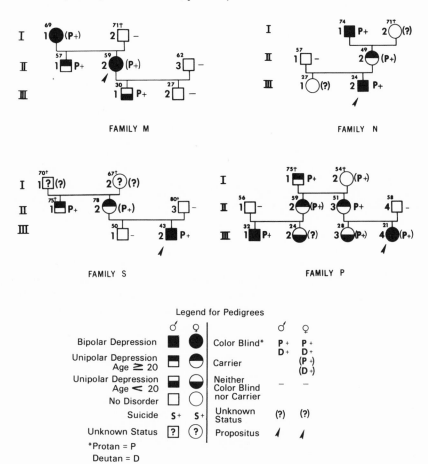

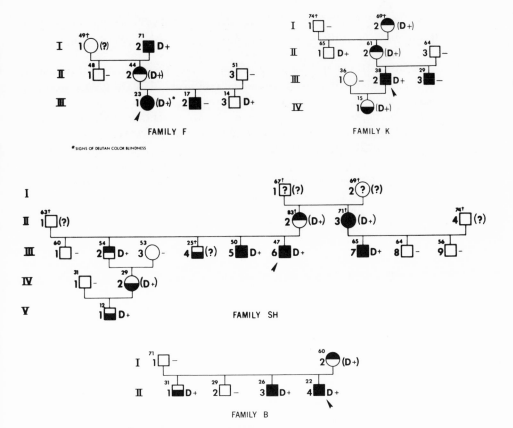

Fig. 2. Pedigrees with Deutan Color Blindness. (Reprinted with permission from Mendlewicz, Fleiss, & Fieve 1972*b*, p. 1626.)

Deutan Color Blindness (Figure 2)

Family F—The proband, III (1), has mildly impaired color vision, having exhibited some signs of deutan color blindness on separate examinations of both eyes. One of her brothers, III (2), has bipolar depression but is not color blind. The proband's maternal grandfather, I (2), is manic-depressive and color blind suggesting that the traits are in coupling. III (1) is counted as a nonrecombinant, and III (2) as a recombinant. The proband's younger brother, III (3), is color blind, but too young to be informative about affective illness. The finding of mild color blindness in the proband, III (1), who is heterozygous for this anomaly is consistent with the Lyon hypothesis of random inactivation of one or the other of the two X chromosomes.

Family K—The proband, III (2), and his brother, III (3), are both manic-depressive, but III (2) is color blind and III (3) is not. Their maternal uncle, II (1), is color blind and well, suggesting that the mother, II (2), has the traits in repulsion. III (2) may therefore be counted as a recombinant and III (3) as a nonrecombinant.

Family SH—In this family pedigree, and in the subsequent pedigree to be described (B), the phase cannot be established but the data can nevertheless be analyzed mathematically for linkage where one takes into account the probability of either phase being present. In addition, we cannot identify recombinants or nonrecombinants. However, from Figure 2 it can be noted that III (2), III (5), III (6), and III (7) are affectively ill and color blind, whereas III (1), III (8), and III (9) are well and not color blind. Thus, although phase cannot be determined, it is apparent that the illness and the marker tend to assort together, as well as the absence of the illness and the absence of the marker.

Family B—The proband, II (4), is manic depressive and color blind. Two of his brothers, II (1) and II (3) are also affectively ill and color blind. Another brother, II (2) is well and not color blind. As previously stated, phase cannot be determined.

Linkage Results

The data were analyzed by the application of the lod score method of Morton (1955) as tabulated by Edwards (1971). The lod scores of the values of the recombination fraction θ between protan color blindness and manic-depressive illness for the four pedigrees in Figure 1, along with the lod scores from one family studied by Winokur and Tanna (1969), are presented in Table 1. Based on the data for all the families the maximum likelihood estimate of the recombination fraction is $\hat{\theta} = 0$. Since the number of families is small this does not imply that the two loci coincide.

The estimated standard error of $\hat{\theta}$ is 0.05, yielding a 95 percent confidence interval for the recombination fraction of $\theta \leq 0.08$. The recombination fraction is therefore significantly different from 0.50, the value for the null hypothesis of no linkage.

Also shown in Table 1 and plotted in Figure 3 are the antilogs of the sums of these scores which represent the odds favoring the given values of θ over 0.50. For the most probable value of θ, namely 0, the odds are 398 to 1.

The lod scores of the values of the recombination fraction between deutan color blindness and manic-depressive illness for the four pedigrees in Figure 2, along with the lod scores from one family studied by Reich et al. (1969), are presented in Table 2. Also shown in Table 2 and plotted in Figure 4 are the antilogs of the sums of these scores. Based on the data for all the families the maximum likelihood estimate of the recombination fraction is $\hat{\theta} = 0.11$ with an upper 95 percent confidence limit of 0.21. The odds favoring this value of θ over 0.50 are 116 to 1.

Table 1. Manic-depressive Illness and Protan Color Blindness

Family	Method of ascertainment	a+d*	b+c*	Recombination fraction θ									
				.0	.05	.10	.15	.20	.25	.30	.35	.40	.45
M	CMBC	2	0	.56	.52	.47	.42	.36	.29	.22	.14	.07	.02
P	Z1	1	0	.30	.28	.26	.23	.20	.18	.15	.11	.08	.04
N	Z1	1	0	.30	.28	.26	.23	.20	.18	.15	.11	.08	.04
S	CMBC	2	0	.56	.52	.47	.42	.36	.29	.22	.14	.07	.02
Alger†	CMBC	3	0	.88	.81	.74	.66	.57	.47	.36	.24	.13	.04
Total				2.60	2.41	2.20	1.96	1.69	1.41	1.10	.74	.43	.16
Antilog = odds favoring θ over 50				398	257	159	91.2	49.0	25.8	12.6	5.5	2.7	1.4

*a = number of sons ill and color blind
b = number of sons ill and not color blind
c = number of sons well and color blind
d = number of sons well and not color blind
†Winokur & Tanna 1969.

247

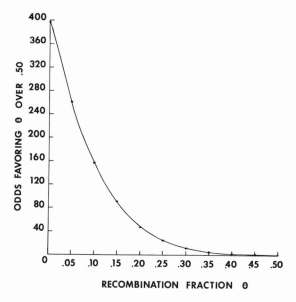

Fig. 3. Odds for Linkage between the Loci of Bipolar Illness and Protan Color Blindness.

Discussion

Studies with three genetic markers by two independent groups of investigators have now shown evidence of X-linkage in manic-depressive illness; in each of these studies the statistical evidence for X-linkage is compelling. In the past there have been many attempts to construct a genetic model for the inheritance of manic-depressive psychosis. The preponderance of females in many surveys suggested the possibility of X-linked inheritance, although most early investigators favored autosomal patterns of inheritance. They tended to ascribe the excess of females to differences in ascertainment and clinical certifiability or to sex-influenced or sex-limited inheritance with hormonal or psychological factors playing a part in the relatively higher penetrance in females. Rosanoff (Rosanoff et al. 1935) was one of those who postulated two dominant genes, one being sex-linked and one autosomal. More recently Winokur and his colleagues (Winokur et al. 1969) have suggested the existence of an X-linked factor in the transmission of bipolar affective disorder, while Slater, Maxwell, and Price (1971) and Perris (1971) have been inclined to favor polygenic inheritance.

While X-borne inheritance may be suggested by pedigree analysis showing certain patterns of inheritance, or by characteristic sex ratios among the children or sibs of probands, the measurement of such linkage

Table 2. Manic-depressive Illness and Deutan Color Blindness

Family	Method of ascertainment	a+d*	b+c*	Recombination fraction θ								
				.05	.10	.15	.20	.25	.30	.35	.40	.45
K	CMBR	1	1	-.76	-.47	-.32	-.21	-.14	-.08	-.05	-.02	0
FR	CGC	1	1	-.83	-.54	-.38	-.26	-.18	-.12	-.07	-.04	-.01
SH	CS2	7	0	1.31	1.16	1.00	.82	.65	.47	.29	.14	.04
B	CS2	4	0	.80	.71	.62	.51	.41	.30	.19	.09	.03
Calvert†	CS2	6	0	1.34	1.20	1.06	.90	.75	.57	.39	.21	.06
Total				1.86	2.06	1.98	1.76	1.49	1.14	.75	.38	.12
Antilog = odds favoring θ over .50				72.4	115	95.5	57.5	30.9	13.8	5.6	2.4	1.3

*a = number of sons ill and color blind
b = number of sons ill and not color blind
c = number of sons well and color blind
d = number of sons well and not color blind
†Reich, Clayton, & Winokur 1969.

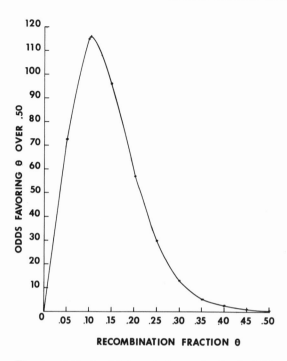

Fig. 4. Odds for Linkage between the Loci of Bipolar Illness and Deutan Color Blindness.

by the statistical method used in the present study is potentially able to distinguish between X-linked and sex-limited types of inheritance and to add to our knowledge of the mapping of the human chromosome. At the present time the number of families studied is limited, but the data provided here are suggestive enough of linkage to provide a strong encouragement for testing more families.

In pedigrees of dominant X-linked transmission an essential feature is the absence of male-to-male transmission. Hoffman (1921), Rüdin (1923), and Slater (1936), however, presented data involving transmission from father to son, while more recently Perris (1968) found father-to-son transmission in bipolar (although not in unipolar) disorder. Winokur et al. (1969) did not note father-son transmission and favored an X-linked factor. It is difficult, if not impossible to evaluate these studies, reported over many decades, from the point of view of diagnostic uniformity, assortative mating, illegitimacy, and carefulness of family investigation.

In our series of 120 bipolar probands, there were 13 pairs of apparent male-to-male transmission. In four of these the mother was also affected, implying that the illness may have been transmitted from the maternal side. Nine pairs, therefore, remained, which were inconsistent with the pattern of X-linked dominant inheritance. Blood tests for ruling out illegitimacy have not yet been completed on this population. The most striking of these pedigrees showing four instances of male-to-male transmission is illustrated in pedigree H (Fig. 5), where assortative mating is most unlikely. Pedigree ST (Fig. 6) in turn, is an example of assortative mating where the mode of inheritance is ambiguous.

In our four cases in which both father and son were affected, and the mother was also affected, transmission could have come from the mother rather than from the father. This may also have been the case in some of the earlier studies; moreover, many of the cases reported in Hoffman's (1921) series represented cyclothymic temperaments rather than diagnosable psychoses and in none of the earlier series were unipolar and bipolar illness separated.

A second essential feature of X-linked dominant transmission is that affected males transmit the trait to all their daughters, whereas affected females transmit the trait to half their sons and half their daughters. In the case of male probands there should be an equal number of affected male and female sibs, but in the case of female probands there should be three times as many affected female sibs as male sibs. Winokur et al. (1969) have found sex ratios consistent with this model for bipolar illness, and Perris (1966) for unipolar illness. In the 120 families from which the present informative pedigrees were taken, a preliminary analysis shows both female and male probands had more affected daughters than sons, but the expectancy rates were in the ratio of almost 3 daughters to 1 son for the female probands and almost 6 daughters to 1 son for the male probands. In the sibships of the female probands there was a slightly higher expectancy for affected sisters than for affected brothers. These sex ratios cannot be explained solely on the basis of X-linked dominant inheritance.

In earlier studies, and in our own (Mendlewicz et al. 1972*b*) the conflicting sex ratios may be attributed to genetic heterogeneity (Mendlewicz, Fieve, Rainer, & Fleiss 1972*a*; Mendlewicz, Fieve, Rainer, & Cataldo 1973) in which the pattern of inheritance, the gene or genes involved, and perhaps the clinical or biochemical nature of the disease may be different in different families. Another explanation may be that of polygenic inheritance in which some or most of the genes are X-linked but in which there are also autosomal genes, the balance of these various genes determining the pattern of inheritance in a particular family. A variation of this model may be an "oligogenic" model in which there are a few genes interacting to produce the illness; this is similar to Rosanoff's early model and Winokur's more recent model.

The maximum likelihood method of linkage analysis is applicable to mendelian genes and has been used successfully in establishing the ten or twelve known linkages on the X chromosome involving such factors as color blindness, G6PD deficiency, Xga blood group, and such clearly defined

252

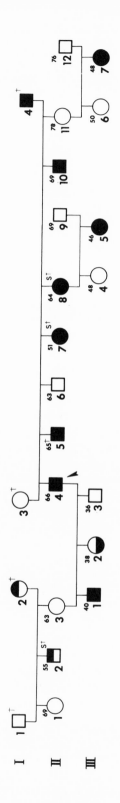

FAMILY H

Fig. 5. Male-to-Male Transmission of Bipolar Illness. Assortative mating is unlikely.

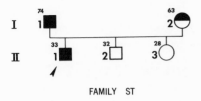

FAMILY ST

Fig. 6. Assortative Mating in Bipolar Illness with Ambiguous Mode of Inheritance.

conditions as hemophilia, ichthyosis, or ocular albinism. Its application to psychiatric illness, in this case manic-depressive disorder, carries with it certain ambiguities. Diagnosis must be made on the basis of clinical criteria as clearly defined as possible, but, nevertheless, is subject to misclassification when the symptoms are not typical. In the present study, the diagnostic criteria have been made as strict as possible to try to meet this problem.

In the present material, the odds reported are lower by a number of orders of magnitude than those that have been found in some of the best established linkages, such as that between G6PD deficiency and color blindness, where the odds for close linkage of about 0.05 are almost 3,000 to 1. In the present data, with low recombination fractions and odds ranging from 100/1 to 400/1, there is evidence of close linkage, at least in some families. These findings, along with our Xg findings (Fieve, Mendlewicz, & Fleiss 1973), indicate that at least in some families the pattern of inheritance is that of an X-linked dominant gene.

To the extent that the findings presented in this paper contribute toward the determination of close linkage between a gene involved in manic-depressive psychosis and known marker genes on the X chromosome, manic-depressive psychosis represents the first instance, aside from certain mental deficiency syndromes, of a demonstration of a mendelian mechanism in psychiatric disorder.

References

Angst, J., & Perris, C. Zur Nosologie endogener Depression. Vergleich der Ergebnisse zweier Untersuchungen. *Archiv für Psychiatrie und Nervenkrankheiten,* 1968, **210,** 373–386.

Edwards, J. H. The analysis of X-linkage. *Annals of Human Genetics,* 1971, **341,** 229–250.

Endicott, J., & Spitzer, R. L. Current and past psychopathology scales: rationale, reliability, and validity. *Archives of General Psychiatry,* 1972, **27,** 678–687.

Farnsworth, D. *The Farnsworth-Munsell 100 hue test.* Baltimore, Md.: Munsell Color Co., 1947.

Fieve, R. R., Mendlewicz, J., & Fleiss, J. L. Manic-depressive illness: linkage with the Xg blood group. *American Journal of Psychiatry*, 1973, **130**, 1355-1359.

Hardy, H., Rand, G., & Rittler, M. C. *Pseudoisochromatic plates* (2nd ed.). American Optical Corporation, 1957.

Hoffman, H. *Die Nachkommenschaft bei endogenen Psychosen.* Berlin: Springer, 1921.

Kallmann, F. J. Genetic principles in manic-depressive psychoses. In P. Hoch & J. Zubin (Eds.), *Depression.* New York: Grune and Stratton, 1954, pp. 1-24.

Kraepelin, E. *Manic-depressive insanity and paranoia.* Edinburgh: E. S. Livinston Publishers, 1921.

Leonhard, K., Korff, I., & Schulz, H. Die Temperamente in den Familien der monopolaren und bipolaren phasischen Psychosen. *Psychiatria Neurologica*, 1962, **143**, 416-434.

Mendlewicz, J., Fieve, R. R., Rainer, J. D., & Cataldo, M. Affective disorder on paternal and maternal sides. Observations in bipolar (manic-depressive) patients with and without a family history. *British Journal of Psychiatry*, 1973, **22**, 31-34.

Mendlewicz, J., Fieve, R. R., Rainer, J. D., & Fleiss, J. L. Manic-depressive illness: a comparative study of patients with and without a family history. *British Journal of Psychiatry*, 1972, **120**, 523-530.(a)

Mendlewicz, J., Fleiss, J. L., & Fieve, R. R. Evidence for X-linkage in manic-depressive illness. *Journal of the American Medical Association*, 1972, **222**, 1624-1627.(b)

McKusick, V. A. *Mendelian inheritance in man: catalogs of autosomal dominant, autosomal recessive and X-linked phenotypes* (3rd ed.). Baltimore: The Johns Hopkins University Press, 1971.

Morton, N. E. Sequential tests for the detection of linkage. *American Journal of Human Genetics*, 1955, **7**, 277-318.

Perris, C. (Ed.) A study of bipolar (manic-depressive) and unipolar recurrent depressive psychoses (I-X). *Acta Psychiatrica Scandinavica*, 1966, **194** (Suppl.).

Perris, C. Genetic transmission of depressive psychoses. *Acta Psychiatrica Scandinavica*, 1968, **203** (Suppl.), 45-52.

Perris, C. Abnormality on paternal and maternal sides: observations in bipolar (manic-depressive) and unipolar depressive psychoses. *British Journal of Psychiatry*, 1971, **118**, 207-210.

Reich, T., Clayton, P. J., & Winokur, G. Family history studies: V. the genetics of mania. *American Journal of Psychiatry*, 1969, **125**, 1358-1369.

Rosanoff, A. J., Handy, L. M., & Plesset, I. R. The etiology of manic-depressive syndromes with special reference to their occurrence in twins. *American Journal of Psychiatry*, 1935, **91**, 725-762.

Rüdin, E. Über Vererbung geistiger Störungen. *Zeitschrift für die gesamte Neurologie und Psychiatrie*, 1923, **81**, 459.

Slater, E. The inheritance of manic-depressive insanity. *Proceedings of the Royal Society of Medicine*, 1936, **29**, 981-990.

Slater, E. *An investigation into psychotic and neurotic twins.* London: University of London, 1951.

Slater, E., Maxwell, J., & Price, J. S. Distribution of ancestral secondary cases in bipolar affective disorders. *British Journal of Psychiatry*, 1971, **118**, 215-218.

Stenstedt, A. A study in manic-depressive psychosis. *Acta Psychiatrica Scandinavica*, 1952, **72** (Suppl.), 1–111.

Tanna, V. L., & Winokur, G. A study of association and linkage of the ABO blood types and primary affective disorder. *British Journal of Psychiatry*, 1968, **114**, 1175–1181.

Winokur, G., & Clayton, P. J. Family history studies: I. two types of affective disorders separated according to genetic and clinical factors. In J. Wortis (Ed.), *Recent advances in biological psychiatry*, Vol. 9. New York: Plenum Press, 1967, pp. 35–50.

Winokur, G., Clayton, P. J., & Reich, T. *Manic-depressive illness*. St. Louis: C. V. Mosby, 1969.

Winokur, G., & Tanna, V. L. Possible role of X-linked dominant factor in manic-depressive disease. *Diseases of the Nervous System*, 1969, **30**, 89–93.

Zerbin-Rüdin, E. Zur Genetik der depressiven Erkrankungen. In H. Hippius & H. Selbach (Eds.), *Das depressive Syndrom*. Munich: Urban and Schwarzenberg, 1969.

GENETIC
STUDIES
IN OTHER
DISORDERS

18 THE TRANSMISSION OF ALCOHOLISM

Theodore Reich, George Winokur,
and J. Mullaney

Introduction

Family studies of alcoholics indicate that alcoholism may be familial and can be transmitted from parent to offspring (Winokur, Reich, Rimmer, & Pitts 1970; Amark 1951). The importance of the role played by genetic factors in the transmission of this disorder is indicated by a comparison of the adopted-out offspring of alcoholics and nonalcoholics. This study showed that alcoholism can be transmitted even if the parent and offspring have separated early in life (Goodwin, Shulsinger, Hermansen, Guze, & Winokur 1973). This finding is supported by a study of half-siblings with alcoholic parents, which showed that having a biological alcoholic parent not present in the home was more of an influence in generating alcoholism than the presence in the home of a genetically unrelated alcoholic parent (Schuckit, Goodwin, & Winokur 1972). Since cultural and social class variables are also important determinants of the form and prevalence of this behavior, it may be safely concluded that both biological and environmental factors are involved in the transmission of alcohol addiction, although the way in which these factors interact and the importance of each is unknown. The present study was undertaken in order to investigate some of the complex relationships among these variables.

The Multifactorial Model of Inheritance seems to be an appropriate way of analyzing data obtained from a study of the families of alcoholics, since it does not require the assumption that the transmission of the illness is entirely genetic. In the analysis described here a modified form of the Multifactorial Model proposed by Falconer (1965) is used throughout. The derivations of the equations for estimating the correlations and their standard errors are found elsewhere (Reich, Wette, & Mullaney 1973;

Theodore Reich, M.D., Associate Professor of Psychiatry, Washington University School of Medicine, St. Louis, Missouri 63110.

George Winokur, M.D., Professor and Head of the Department of Psychiatry, University of Iowa, College of Medicine, Iowa City, Iowa 52240.

J. Mullaney, M.D., Research Assistant in Computer Sciences in Psychiatry, Washington University School of Medicine, St. Louis, Missouri 63110.

This paper is supported in part by U.S.P.H.S. Grants AA-00209, MH-13002 and MH-07081

Reich, James, & Morris 1972). In this paper, correlations which estimate the liabilities to develop alcoholism are made in a white and in a black population, in order to measure the degree to which the disorder is familial. This is done by comparing the population prevalence of the disorder with the prevalence among the first-degree relatives of affected individuals. Then the correlations within and between the members of each sex are calculated to assess the phenotypic correlation between alcoholism in males and females. Finally, several models of inheritance which compare the role of transmitted factors and individual environmental variation in determining the difference in the prevalence of male and female alcoholism are fitted to the data.

The Multifactorial Model of Inheritance

The multifactorial model of inheritance requires the following assumptions: (1) All genetic and environmental causes of the disorder may be combined into a single continuous variable termed the *liability*. (2) Individuals whose liability exceeds a threshold are affected. (3) The distribution of liability in the general population is normal or can be transformed to a normal distribution. (4) Genes which are relevant to the etiology of the disease are each of small effect in relation to the total variation, and are additive. (5) Environmental factors are due to many events whose effects are additive. (6) The model does not assume the absence of environmental effects common to relatives.

The distribution of the liability in the general population and the relatives of affected individuals is displayed in Figure 1.

If the assumptions of the model are valid then,

$$r = \frac{X_p - X_R \sqrt{1 - (X_p^2 - X_R^2)(1 - \dfrac{X_p}{a})}}{a + X_R^2(a - X_p)} \tag{1}$$

where r = the phenotypic correlation of liability between relatives,

K_p = the population prevalence of the disorder,

K_R = the prevalence of the disorder in the relatives of affected probands,

X_p and X_R are the normal deviates for K_p and K_R,

a = the mean deviation of affected individuals from the population mean;

$$a = \frac{1}{\sqrt{2\pi}} \frac{e^{-\frac{1}{2} X_p^2}}{K_p}$$

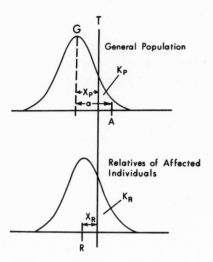

Fig. 1. The Multifactorial Model of Inheritance. Distribution of liability in the general population (*above*) and in relatives of affected individuals (*below*). G = mean liability of population; R = mean liability of relatives; A = mean liability of probands; K_p = affected proportion of population; K_R = affected proportion of relatives; X_p, X_R = normal deviates for K_p, K_R; a = mean deviation of probands from population mean.

The correlation (r), is an estimate of the degree to which the disorder clusters in families, and hence is a measure of the magnitude of between-family sources of variation in relation to the total variation. When the correlation is measured from first-degree relatives, the degree of genetic determination i.e., the heritibility (h^2) is,

$$h^2 = 2r$$

This estimate is valid only if the environments of the relatives are uncorrelated.

The correlation (r) takes the population prevalence into account, enabling different populations and different disorders to be compared on the same scale in terms of the importance of between-family or transmissible factors in the disease process. The value of the correlation is 0 when the proportion of affected relatives equals the population prevalence and is 1 when all the relatives are affected. If the disease is entirely due to additive genetic factors, then the value of the correlation between first-degree relatives is expected to be 0.5. If correlations greater than 0.5 are found, then factors other than additive genes are responsible for a portion of the similarity between relatives.

Alcoholism is about four to five times as common in males as females; accordingly, the correlations between male probands and male relatives (r_{mm}), and between female probands and female relatives (r_{ff}), are estimated separately (Equation 1). The cross correlations, that is, the correlations between probands and relatives of opposite sexes can also be estimated.

$$r_{mf} = \frac{X_{pm} - X_{Rmf} \sqrt{1 - (X_{pf}^2 - X_{Rmf}^2)(1 - \dfrac{X_{pm}}{a_m})}}{a_m + X_{Rmf}^2(a_m - X_{pm})} \tag{2}$$

where r_{mf} = the correlation derived from male probands and female relatives, also

$$r_{fm} = \frac{X_{pm} - X_{Rfm} \sqrt{1 - (X_{pm}^2 - X_{Rfm}^2)(1 - \dfrac{X_{pf}}{a_f})}}{a_f + X_{Rfm}^2(a_f - X_{pf})} \tag{3}$$

where r_{fm} = the correlation derived from female probands and male relatives. The subscripts m, f, refer to male and female probands and relatives. Where two subscripts appear, the first indicates the sex of the proband and the second the sex of the relative. The estimates of r computed separately for each sex require the assumption that the liability to develop the disorder is normally distributed in each sex.

If the assumptions of the model are valid, r_{mf} is not expected to differ from r_{fm}. The four values of r obtained by classifying probands and relatives according to sex are used to estimate the phenotypic correlation (r_p) between alcoholism in males and females. This correlation is a measure of the independence of the trait in the two sexes. A correlation of 1 indicates that the disorders are not independent and that the trait is not transmitted separately in the families of male and female probands. A correlation of 0 indicates that the disorder is transmitted independently. The standard error of the phenotypic correlation is usually quite large and this statement cannot be made with great precision. The phenotpic correlation r_p is,

$$r_p = \frac{r_{mf}}{\sqrt{r_{mm}r_{ff}}} = \frac{r_{fm}}{\sqrt{r_{mm}r_{ff}}} \tag{4}$$

A method for the estimation of the standard error of r_p, assuming that the environments of the probands and relatives are uncorrelated, is given by Smith (Smith, Falconer, & Duncan 1972).

The environmental and cultural phenomena which are relevant to alcoholism can be divided into those which increase the resemblance among relatives and those which are not dependent on the family to which an individual belongs. The former serve to increase the between-family variation and hence increase the estimate of the correlation between relatives. These factors seem to play a greater role in the transmission of alcoholism between male probands and male relatives. The effect of the latter, i.e., individual nonfamilial environmental variation, is to decrease the resemblance between relatives and hence to decrease the correlation which measures it. This sort of variation seems to play a greater role in the variation of the liability to develop alcoholism in women.

Methods

Alcoholic probands were drawn from consecutive admissions to two psychiatric hospitals, and an attempt was made to interview all first-degree relatives living in the St. Louis area. The probands and their relatives were examined by means of an extensive structured interview. Alcoholism was diagnosed when significant social or medical problems were found which were the result of alcoholic abuse. There were 259 probands, of whom 194 (115 male) were white and 63 (39 male) were black. Because alcoholic women patients are less common than men, a greater proportion of hospitalized alcoholic women were accepted as probands. A total of 510 first-degree relatives were interviewed.

Population prevalences were derived from a national survey of problem drinking by Cahalan (1970). Age-specific rates from the national survey were used to construct prevalences of alcoholism in a population whose age was comparable to that of the relatives. Only those problem drinkers in the Cahalan survey who suffered major social or medical problems as a result of alcohol abuse were counted. Since age-specific rates for blacks could not be found, values from the lower socioeconomic group in the national survey were used. The age distribution of the relatives of male and female probands was similar, and a single population prevalence for each sex was calculated for comparison with the male and female relatives. The prevalence of alcoholism was 11.4 percent among white males, 2.9 percent among white females, 17 percent among black males and 3.1 percent among black females.

Results

The population prevalences and the proportion of affected relatives comparable to them are displayed in Table 1.

The prevalences among the relatives displayed in Table 1 are corrected for the excess of female probands in the hospitalized sample when compared with the population prevalence of female alcoholism and assume

Table 1. Correlations in Liability between First-Degree Relatives in a White and
Black Population, Assuming the Multifactorial Model of Inheritance

	White population	Black population
Population prevalence	7.2%	10.0%
Prevalence in first-degree relatives	20.7%	31.9%
Number of affected relatives	66	38
Correlation	0.36 ± 0.05	0.49 ± 0.07

an equal number of male and female relatives. The correlations of
0.36 ± 0.05 among whites and 0.49 ± 0.07 among blacks indicate that
between-family sources of variation, either genetic or cultural, for each
race are extremely important features of this disorder. Because of the findings in the adoption and half-sibs studies cited earlier it seems likely
that these correlations represent, in part, genetic transmission of the
liability to develop alcoholism.

Sex Effect in Alcoholism

The nature of the great preponderance of male alcoholics can be explored
by estimating the correlations in liability between members of the same
sex and between probands and relatives of opposite sex. The prevalence of
alcoholism in the male and female relatives of white alcoholic probands
is displayed in Table 2a. The correlations and cross correlations computed
from these data and from the population prevalences for alcoholism in
white males and females is shown in Table 2b.

The correlation between males is 0.53 ± 0.05 and between females,
0.18 ± 0.10, a significant difference. This finding indicates that non-
transmissible environmental factors play a greater role in the variance
of female than male alcoholism. The cross-correlations do not differ

Table 2a. Prevalence of Alcoholism among the Male and Female First-Degree
Relatives of White Male and Female Probands

		First-degree relatives			
		Male % affected	Total number	Female % affected	Total number
Probands	Male	36.0	111	6.4	156
	Female	31.6	38	6.7	60

K_{pm} = 11.4% K_{pf} = 2.9%

Table 2b. Correlations in Liability between White Male and Female Probands
and Their Like-sexed and Opposite-sexed Relatives

| | | First-degree relatives | |
		Male	Female
Probands	Male	0.53 ± 0.05	0.24 ± 0.07
	Female	0.33 ± 0.09	0.18 ± 0.10

$r_{pfm} = 1.06 \pm 0.19, r_{pmf} = 0.78 \pm 0.12$

greatly, as would be expected if the assumptions of the multifactorial
model are valid. The phenotypic correlation between male and female
alcoholism is (1.06 ± 0.19) when estimated for female probands and male
relatives, and (0.78 ± 0.12) when estimated for male probands and female
relatives. These phenotypic correlations do not differ significantly from
unity and indicate that male and female alcoholism are not independent
and do not run in separate families. If the phenotypic correlation is largely
due to genetic factors, then the genotypes of male and female alcoholics are expected to be drawn from the same population.

Data obtained from black probands and their relatives were analyzed
in the same fashion as above. The prevalence of alcoholism among the
male and female relatives of black probands is given in Table 3a.

As in the case of alcoholism among whites the correlation between
black males is greater than that between females, again suggesting that
individual nonfamilial environmental factors are more important features of female alcoholism. The cross-correlations do not differ greatly,
supporting the validity of the initial assumptions. Among blacks the phenotypic correlations are rather low ($r_{pmf} = 0.85 \pm 0.2$ and $r_{pfm} = 0.78 \pm 0.2$).
However, the standard errors are large and the values of r_p do not differ
significantly from unity. It seems likely that for blacks, male and female

Table 3a. Prevalence of Alcoholism Among the Male and Female First-Degree
Relatives of Black Male and Female Probands

| | | First-degree relatives | | | |
		Male % affected	Total number	Female % affected	Total number
Probands	Male	59.5	37	8.6	58
	Female	44.4	18	10.3	29

$K_{pm} = 17\%, \quad K_{pf} = 3.1\%$

Table 3b. Correlations in Liability Between Black Male and Female Probands
and Their Like-sexed and Opposite-sexed Relatives

| | | First-degree relatives | |
		Male	Female
Probands	Male	0.76 ± 0.06	0.40 ± 0.09
	Female	0.36 ± 0.10	0.29 ± 0.13

$r_{pfm} = 0.78 \pm 0.20, r_{pmf} = 0.85 \pm 0.20$

alcoholism are not independent, or if they are independent, are transmitted by factors which largely overlap.

The Difference between Alcoholism in Males and Females

The preceding analysis suggests that the between-family factors relating
to alcoholism largely overlap when the families of male and female
probands are compared, and that individual environmental factors are
more important features of the variance of alcoholism in women. Further
understanding of the differences between these groups can be obtained
by comparing the data with several models of inheritance and seeing
which model might fit the observations.

The Two-Threshold Model   This model assumes that variation in
the liability to develop alcoholism in males and females is made up of the
same proportions of genetic and environmental variance and that the
smaller proportion of affected females is due to the threshold for females
being more deviant than that of males. Under these conditions the genotypes of affected females are expected to be more deviant than those of
affected males. Such a mechanism is believed to occur in congenital py-
loric stenosis, which is also less common in females. If transmission of
alcoholism were largely due to cultural factors, then the model would
require that affected females be more deviant in this regard.

By examining Figure 2 it can be seen that in the two-threshold model
the mean liability of female probands (A_f) is more deviant than that of
male probands (A_m). The relatives of female probands should also be more
deviant, and this model predicts that the prevalence of affected male and
female relatives of female probands is greater than that of male probands.
The model also requires that the values of r_{mm}, r_{ff}, and r_{mf}, and r_{fm},
be equal. A diagrammatic representation of the two-threshold model is
given in Figure 2.

The model was tested by choosing a wide range of values of K_{pm}, K_{pf},
and r and using these parameters to compute expected values for K_{Rmm},

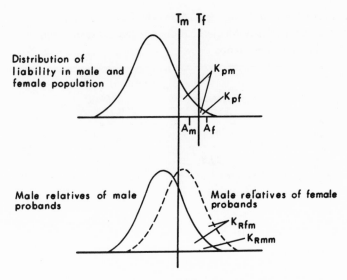

Fig. 2. The Two-Threshold Model, showing the distribution of liability in the male and female population (*above*) and in the male relatives of male and female probands (*below*) T_m = male threshold, T_f = female threshold. K_{pm}, K_{pf} = proportion of affected males and females in general population. A_m, A_f = mean liability of affected males and females. K_{Rfm}, K_{Rmm} = proportion of affected male relatives of male and female probands.

K_{Rff}, K_{Rmf}, and K_{Rfm}. These expected values of K_R were compared with the observed values by means of a chi-square goodness-of-fit test. The parameters used to produce the expected values of K_R were iterated until the chi-square was minimized. The procedure was done separately for white and black populations. The expected proportions of affected first-degree relatives and the expected population prevalences assuming the two-threshold model are given in Tables 4a and 4b.

The chi-square goodness of fit of the model to the data was 6.99 ($0.10 > p > 0.05$) for whites and 8.5 ($0.05 > p > 0.025$) for blacks. Degree of freedom was 3 in each case. The model does not fit the data well and is probably not a valid description of the difference between male and female alcoholics.

The Environmental Model   The environmental model assumes that the difference between male and female alcoholics is entirely due to individual nonfamilial environmental causes of variation, and that a certain proportion of women are not at risk with respect to alcoholism. It also assumes that the genotypic deviation of male and female alcoholics from the population mean is equal, and males and females are equally potent in transmitting the disorder. If "between-family" cultural factors are important causes of the familial nature of the disorder, or its trans-

Table 4a. Expected Prevalence of Alcoholism among the First-Degree Relatives
of White Male and Female Alcoholic Probands Assuming the Two-
Threshold Model of Inheritance

| | | First-degree relatives | |
		Male % affected	Female % affected
Probands	Male	28.1	7.8
	Female	37.8	12.4

K_{pm} = 11.9% ± 1% K_{pf} = 2.5% ± 0.5% r = 0.38 ± 0.05 X_3^2 = 6.99 (0.10>p>0.05)

mission, then they are expected to be equally present in the families of
male and female alcoholics. The nature of the individual environmental
factors are not specified by the model. One possibility, for example, would
be that the degree of exposure sufficient to develop alcoholism occurs less
often in women. Alternatively, women may conceal their alcoholic problems better than males. The latter possibility does not seem very likely,
as the more severe grades of alcoholism are difficult to conceal and are also
less common in women. The distribution of the liability to develop alcoholism in the general population is displayed in Figure 3.

By examining Figure 3 it can be seen that the difference in mean
deviation of affected males and females from the population mean is
small. The proportion of male and female relatives affected with alcoholism in the two groups is therefore not expected to differ, even though
affected females are less common. The acceptability of this model was
tested by comparing the proportion of affected male relatives of male and
female probands and the proportion of affected female relatives of male
and female probands. The chi-square value for whites is 0.25 with 2 d.f.

Table 4b. Expected Prevalence among the First-Degree Relatives of Black
Alcoholic Probands Assuming the Two-Threshold Model of
Inheritance

| | | First-degree relatives | |
		Male % affected	Female % affected
Probands	Male	41.4	9.5
	Female	59.3	19.3

K_{pm} = 17.4% ± 2.9% K_{pf} = 2.8% ± 0.7% r = 0.50 ± 0.06 X_3^2 = 8.5 (0.05>p>0.025)

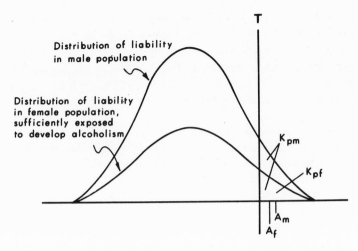

Fig. 3. The Environmental Model, showing the distribution of liability in the male and female population, assuming a proportion of the female population is not at risk. T = threshold, A_f = mean liability of affected females, A_m = mean liability of affected males, K_{pm} = prevalence in males, K_{pf} = prevalence in females.

(0.90 > p > 0.75) and for blacks is 1.17 with 2 d.f. (0.75 > p > 0.50). This model adequately explains the differences between the population prevalences of male and female alcoholism and suggests that these differences can be entirely explained by individual environmental sources of variation.

Discussion

The correlation in the liability to develop alcoholism cannot be used to estimate the degree to genetic determination of this trait accurately, since it may include nongenetic causes of the similarity among relatives. Some measure of the importance of the transmissible factors in determining the presence of the disorder can be made by comparing the correlations for alcoholism with those for other complex disorders in which environmental factors may play an important role. The correlation between first-degree relatives for diabetes is 0.30, for peptic ulcer is 0.18, and for essential hypertension is 0.20 (Falconer 1965; Smith et al. 1972; Hamilton, Pickering, Fraser Roberts, and Lowry 1954). The correlation of 0.36 ± 0.55 and 0.49 ± 0.07 for alcoholism in white and black populations is higher than these and suggests that between-family factors, which include genetic transmission, account for a large portion of the total variance.

If the transmission of alcoholism is entirely due to additive genes then the expected value of r is 0.5. The correlation between males in both the white and black families exceeds this value (see Tables 2b and 3b). One reason for this finding might be that nonadditive genetic factors, such as dominance, are present. This can be ruled out, as the proportion of affected fathers and brothers is nearly equal. Another possibility is that the population prevalences are an underestimate. If the population prevalence of alcoholism in black males is 30 percent then the value of r_{mm} is 0.63, which still exceeds 0.5. The most likely explanation is that cultural factors are involved in the transmission of alcoholism between males. Since the phenotypic correlation between alcoholism in males and females approaches unity, cultural factors are also expected to be features of transmission between females and between opposite sexed relatives.

The estimates of the correlations between relatives, the phenotypic correlations, and the fitting of the two-threshold model, all rest on the assumptions of multifactorial inheritance. These assumptions cannot be tested by data presented here. The environmental model is more general in its conclusions, since it only assumes that if two groups are equally potent in transmitting a disorder, the genotypes of the two groups do not differ. If alcoholism is transmitted by a modified single-gene mechanism, the proportion of heterozygotes and homozygotes among male and female alcoholics would be expected to be equal in order to explain the observation that an equal proportion of their relatives are affected. The conclusion that the decreased prevalence of alcoholism in females can be explained by individual environmental variation would be arrived at for the modified single-gene model of inheritance.

The expected distribution of cases in the general population and among the first-degree relatives of male and female probands in the environmental model is the same for single-gene or multifactorial inheritance. Accordingly the observation that this model fits the data is strengthened, since it appears valid for several of the models currently being used to explain the transmission of the common familial diseases.

Summary and Conclusions

The multifactorial model of disease transmission appears to be a powerful device for the analyses of population and family data without requiring genetic assumptions which have led to such acrimony in the past. With respect to alcoholism our results show that:

a. Among black or white populations alcoholism is familial to a large degree.
b. Alcoholism in males and alcoholism in females is not transmitted independently.
c. The correlations between males are greater than expected if alcoholism is entirely due to the effects of additive genes, implying that cultural transmission is in part responsible for the similarity between

male, between female, and between opposite sexed first-degree relatives.

d. In spite of the phenotypic difference between male and female alcoholism (as shown by the different population prevalences), the genotypes of male and female alcoholics do not differ; this suggests that the large sex effect in this disorder can be explained by individual, nonfamilial, environmental sources of variation. Analysis of the data using other models of inheritance, such as modified single-gene transmission, leads to a similar conclusion.

References

Amark, C. A study in alcoholism. *Acta Psychiatrica Scandinavica*, 1951, **26**, Supplementum 70.

Cahalan, D. *Problem drinkers*. San Francisco: Jossey-Bass, 1970.

Falconer, D. S. The inheritance of liability to certain diseases estimated from the incidence among relatives. *Annals of Human Genetics*, 1965, **29**, 51.

Goodwin, D., Shulsinger, F., Hermansen, L., Guze, S., & Winokur, G. Alcohol problems in adoptees raised apart from biological parents. *Archives of General Psychiatry*, 1973, **28**, 238-243.

Hamilton, M., Pickering, G. W., Fraser Roberts, J. A., & Lowry, G. S. C. The aetiology of essential hypertension 4. The role of inheritance. *Clinical Science*, **13**, 11.

Reich, T., James, J. H., & Morris, C. A. The use of multiple thresholds in determining the mode of transmission of semi-continuous traits. *Annals of Human Genetics*, 1972, **36**, 163.

Reich, T., Wette, H., & Mullaney, J. The analysis of heterogeneity of multifactorial traits. 1973, in preparation.

Schuckit, M., Goodwin, D. W., & Winokur, G. The half-sibling approach in a genetic study of alcoholism. In M. Roff, L. N. Robins, & M. Pollack (Eds.), *Life history research in psychopathology*, Vol. 2. Minneapolis: University of Minnesota Press, 1972, pp. 120-127.

Smith, C., Falconer, D. S., & Duncan, J. P. A statistical and genetical study of diabetes II: heritiability of liability. *Annals of Human Genetics*, 1972, **35**, 281.

Winokur, G., Reich, T., Rimmer, J., & Pitts, F. Alcoholism, III: Psychiatric and familial psychiatric illness in 259 alcoholic probands. *Archives of General Psychiatry*, 1970, **23**, 104-111.

19

GENETIC STUDIES OF HYPERACTIVE CHILDREN: PSYCHIATRIC ILLNESS IN BIOLOGIC AND ADOPTING PARENTS

Dennis P. Cantwell

The hyperactive child syndrome (HACS) was first described by a German physician over one hundred years ago (Hoffmann 1845). Since then several authors (Stewart, Pitts, Craig, & Dieruf 1966; Clements 1966) have delineated a syndrome that begins early in life, is more common in boys, and is characterized by a symptom pattern of overactivity, distractibility, impulsiveness, and excitability. Other difficulties for hyperactive children are peer relations, discipline problems, and specific learning problems. Terms like "minimal cerebral dysfunction" and "minimal brain damage" used to describe this syndrome reflect the assumption of many clinicians that its cause is some type of brain damage. However, only a small minority of hyperactive children show "hard" evidence of definite brain injury (Stewart et al. 1966; Chess 1960; Minde & Webb 1968).

In contrast to other psychiatric disorders of childhood there has been little investigation of familial factors in the hyperactive child syndrome, probably due to the assumption of some "organic" etiology. What little work has been done suggests this may be a fruitful area for investigation.

Two studies (Morrison & Stewart 1971; Cantwell 1972) of biologic parents of hyperactive children have revealed increased prevalence rates for alcoholism, sociopathy, and hysteria. One of these studies (Cantwell 1972) also reported a high prevalence for these same psychiatric disorders in the biologic second-degree relatives of hyperactive children. In both studies it was further noted that hyperactivity also occurred more often in the biologic first- and second-degree relatives of hyperactive children than

Dennis P. Cantwell, M.D., Director of Residency Training in Child Psychiatry, University of California, Los Angeles.

The author wishes to thank Judy Williams and Judy Klusza for statistical assistance and Fran Miller for clerical assistance. This publication was supported in part by Grant MCH-927. Computing assistance was obtained from the Health Sciences Computing Facility, UCLA, sponsored by NIH Special Research Resources Grant RR-3.

in the relatives of control children. Moreover, a significant number of these "grown up" hyperactive children were given diagnoses of alcoholism, sociopathy, and hysteria as adults.

These data suggested two hypotheses: (1) There is a familial relationship between HACS and the three adult psychiatric disorders: alcoholism, sociopathy, and hysteria. (2) The HACS is a familial disorder that passes from generation to generation. The data do not explain whether the familial relationship is a genetic or environmental one nor whether the mechanism of transmission is genetic or environmental.

To test the hypothesis that the relationship between HACS and the three adult psychiatric disorders is a genetic one and that the HACS is genetically transmitted, a systematic psychiatric examination of the nonbiologic parents of adopted hyperactive children was carried out. If the nonbiologic parents and their extended families do not show the same increased prevalence rates for hyperactivity and other psychiatric disorders found in the biologic parents, then an argument could be made for a genetic component operating in the HACS. This paper reports the results of that study.

Method

The biologic probands were 50 hyperactive boys selected at random from more than 300 children referred to the author for evaluation of learning and behavior problems. Inclusion criteria were (1) white male between 6 and 11 years of age, (2) currently attending school, (3) tested normal vision and hearing, (4) IQ of 80 or above (Wechsler Intelligence Scale for Children Full Scale), (5) from an intact family with both biologic parents living in the home, (6) diagnosed as suffering from the hyperactive child syndrome by the criteria of Stewart et al. (1966), which require definite evidence of hyperactivity and distractibility and the presence of 6 of the symptoms found to be most characteristic of the syndrome.

The 39 adopted hyperactive probands included a group referred to the author and a group garnered from the practice of several private physicians in the Southern California area who specialize in the treatment of hyperactive children.* They met the same criteria as the biologic hyperactive probands with the exception of criterion number (5). This criterion for the adopted group was that they have had no contact with their biologic parents after 1 month of age and live in an intact family with both *adopting* parents living in the home.

The 50 control children were selected from a pediatrician's practice. They were screened by the pediatrician to assure that there was no

*The help of Milton Borenstein, M.D. in locating patients and controls is gratefully acknowledged.

hyperactivity in their family and that they came from an intact family with both biologic parents in the home. The three groups were well matched for age, race, sex, and social class.

Each parent was interviewed separately, using an interview described in previous publications (Guze, Tuason, Gatfield, Stewart, & Picken 1962). This interview was systematic and structured. It included a history of current and past illnesses and injuries, a description of all hospitalizations and operations, and a detailed symptom inventory designed to elicit the manifestations of anxiety neurosis, hysteria, obsessional neurosis, schizophrenia, primary affective disorder, organic brain syndrome, drug abuse, alcoholism, sociopathy, and homosexuality. In addition, a detailed family history of psychiatric difficulties and a detailed history of the parental home experience were obtained. The interview also included sections dealing with school history, job history, marital history, military experiences, and police troubles. Specific inquiry was also made about suicide attempts. A mental status examination concluded the interview.

The clinical picture presented by each parent was categorized in terms of syndromes which have been systematically described (Feighner, Robins, Guze, Woodruff, Winokur, & Munoz 1972): alcoholism, hysteria and probable hysteria, sociopathy, primary affective disorder, and organic brain syndrome. An undiagnosed group contained those subjects who were felt to be psychiatrically ill, but whose symptoms did not meet the specific checklist criteria for the diagnoses listed above or other psychiatric diagnoses. In addition, an attempt was made to characterize a parent as having been a hyperactive child himself, if he had demonstrated symptoms in both of the following areas during childhood: (1) learning difficulties, short attention span, distractibility, and poor concentration; (2) hyperactivity, impulsivity, recklessness, and aggressivity. As part of the family history elicited from the interviewed parents, information was obtained about first-degree relatives of the person being interviewed. This information was concerned primarily with hyperactivity, alcoholism, antisocial behavior, affective episodes, schizophrenia, suicide attempts, successful suicides, and psychiatric hospitalization. The only psychiatric diagnoses attempted for a noninterviewed relative based upon the family history obtained from the interviewed parents were hyperactivity, hysteria, alcoholism, schizophrenia, primary affective disorder, organic brain syndrome, and sociopathy. These were global diagnoses using the same general criteria as were used for the diagnoses for the interviewed parents. Statistical analysis was by chi square.

Results

The psychiatric diagnoses for the interviewed parents are presented in Table 1. The data indicate that most of the interviewed parents in the con-

Table 1. Individual Psychiatric Diagnoses of Interviewed Parents (in percent)

	Biologic		Adopted		Control	
	Mothers (N=50)	Fathers (N=50)	Mothers (N=39)	Fathers (N=39)	Mothers (N=50)	Fathers (N=50)
Alcoholism	8[+]	30[c]	0	5	0	14
Sociopathy	0	16[b+]	0	0	0	4
Hysteria	12[a++]	0	0	0	0	0
Probable hysteria	4	0	3	0	2	0
Primary affective disorder						
Unipolar type	8	4	5	5	8	4
Bipolar type	2	0	3	0	2	0
Undiagnosed	2	4	5	5	0	2
Total psychiatrically ill	36[a+++]	54[c+++]	15	15	12	24

a = p < .05   Biologic vs. adopted   + = p < .05   Biologic vs. controls
b = p < .025   Biologic vs. adopted   ++ = p < .025   Biologic vs. controls
c = p < .005   Biologic vs. adopted   +++ = p < .005   Biologic vs. controls

trol group and in the adopted group were free of any psychiatric illness, whereas nearly half of the biologic parents of the hyperactive children had some psychiatric diagnosis (p < .005). The specific differences between the groups are in the greater prevalence of alcoholism, sociopathy, hysteria, and probable hysteria in the biologic parents of hyperactive children.

The clinical diagnoses of the noninterviewed relatives are summarized in Table 2. These findings from the family histories of the interviewed parents are very similar to the comparable ones from the personal interviews. In particular, they confirm the high prevalence of alcoholism, sociopathy, and hysteria among the biologic relatives of the hyperactive chil-

Table 2. Individual Psychiatric Diagnoses of Noninterviewed Relatives
(in percent)

	Biologic		Adopted		Controls	
	Females (N=263)	Males (N=251)	Females (N=176)	Males (N=218)	Females (N=245)	Males (N=256)
Alcoholism	2[+]	20[c+++]	0	3	0	5
Sociopathy	0	12[c+++]	0	1	0	1
Hysteria	8[c+++]	0	1	0	0	0
Primary affective disorder						
Unipolar type	0	1	1	1	1	0
Bipolar type	0	0	0	0	0	0
Organic brain syndrome	0	1	0	0	1	0

a = p < .05   Biologic vs. adopted   + = p < .05   Biologic vs. controls
b = p < .025   Biologic vs. adopted   ++ = p < .025   Biologic vs. controls
c = p < .005   Biologic vs. adopted   +++ = p < .005   Biologic vs. controls

Table 3. Relatives Diagnosed as Hyperactive

	Biologic		Adopted		Controls	
	Total number	Percent hyperactive	Total number	Percent hyperactive	Total number	Percent hyperactive
Mothers	50	4	39	0	50	0
Fathers	50	16[a++]	39	3	50	2
Aunts	163	0	98	0	145	0
Uncles	151	15[c+++]	140	0.7	156	0
First cousins						
Female	307	2[a++]	211	0	282	0
Male	251	12[c+++]	267	1	248	2
Total male relatives	452	12[c+++]	446	1	454	1
Total relatives	966	6.3[c+++]	794	0.6	931	0.6

a = p < .05   Biologic vs. adopted   + = p < .05   Biologic vs. controls
b = p < .025   Biologic vs. adopted   ++ = p < .025   Biologic vs. controls
c = p < .005   Biologic vs. adopted   +++ = p < .005   Biologic vs. controls

dren. These conditions are not found in excess in the nonbiologic relatives of the adopted hyperactive children.

The data concerning hyperactive relatives are presented in Table 3. Of note is the high incidence of hyperactivity in all the biologic male relatives (fathers, uncles, and first cousins) of the hyperactive children. It could be argued that this is an artifact produced by parents of hyperactive children being "sensitized" to the diagnosis and too quick to recognize the symptoms in themselves and in their relatives. However, the low incidence of the syndrome in the nonbiologic relatives of the adopted hyperactive children does not support this argument and adds weight to the idea of a genetic component operating in the HACS.

Comment

The data from this study suggest that a significant percentage of the biologic parents of hyperactive children are psychiatrically ill. Systematic psychiatric examination of the parents revealed high prevalence rates for alcoholism, sociopathy, and hysteria. Family history data elicited from the interviewed parents confirmed the high prevalence of alcoholism, sociopathy, and hysteria among the biologic second-degree relatives of hyperactive children. Increased prevalence rates for these conditions were not found in the nonbiologic first- and second-degree relatives of adopted hyperactive children.

The familial association of alcoholism, sociopathy, and hysteria has been noted in a number of previous studies (Guze et al. 1962; Guze, Wolfgram, McKinney, & Cantwell 1967; Perley & Guze 1962; Robins 1966; Arkonac & Guze 1963; Guze, Goodwin, & Crane 1970; Woerner & Guze 1968; Forrest

1967; Cloninger & Guze 1970). Several authors have presented data suggesting that this association may be, at least in part, a genetic one (Crowe 1972; Rutter 1966). The data from two previous studies (Morrison & Stewart 1971; Cantwell 1972) suggest a familial relationship between HACS and these same three adult psychiatric disorders—alcoholism, sociopathy, and hysteria. Several environmental mechanisms could explain this association. Some authors (Rutter 1966) feel that parental mental illness can produce psychiatric symptoms in children through some nonspecific environmental influence. Learning theorists might go further and argue that a parent could "teach" his child to be hyperactive through selective reinforcement of certain behaviors, or through modeling. A direction of effect from child to parent could also be hypothesized; that is, parents demonstrate psychopathology as a result of living with a deviant child. However, the relative absence of psychopathology in the parents of the adopted hyperactive group does not lend support to a purely environmental explanation for the association of HACS and psychiatric disorders in the parents. The data from this study then tend to suggest that the familial relationship between HACS and alcoholism, sociopathy, and hysteria in parents is a genetic one.

This study provides even stronger evidence for the hypothesis that there is a genetic transmission of the HACS from generation to generation. Table 3 clearly shows that the HACS is found to a much greater degree in the biologic first- and second-degree relatives of hyperactive children than in the adopted relatives. The prevalence rates for the syndrome found in the adopted relatives is no greater than that found for the relatives in the control group and are less than prevalence rates for the syndrome in the general population (O'Malley & Eisenberg 1973). These data clearly favor a genetic mechanism operating in the transmission of the syndrome.

Though these data are strongly suggestive of genetic factors operating in the HACS, several notes of caution are necessary. The first problem is the question of the heterogeneity of the HACS when defined behaviorally as in this paper (Fish 1971). Studies of results of stimulant drug treatment of these children show that there are some who respond dramatically to medication and others who are made worse (Fish 1971). Follow-up studies indicate that all hyperactive children do not have a uniform course (Menkes, Rowe, & Menkes 1967; Weiss, Minde, Werry, Douglas, & Nemeth 1971; Mendleson, Johnson, & Stewart 1971). Neurologic and neurophysiologic studies also indicate that these children are a heterogeneous group (Satterfield 1973). It is likely then that if there is a genetic component to the syndrome, it is operating in one subgroup of these children or there are several genetically distinct subgroups. Future investigation should include family studies of various subgroups—for example, "drug responsive" versus "non-drug responsive" children.

Since the interviewer was not blind to the identity of the parents in the biological, adopted, and control groups, it is possible that interviewer bias could affect the results. The use of a structured interview, operationally defined criteria for scoring a symptom as positive, and operationally defined diagnostic criteria for psychiatric illness (Feighner 1972) somewhat compensates for this defect. However, future investigations should attempt blind collection of data (which may not be entirely possible in studies of this nature).

Finally, since no information was available on the *biologic* parents of the adopted hyperactive children, we can say nothing about the prevalence rates for hyperactivity, alcoholism, sociopathy, and hysteria in that population. Comparison with the prevalence rates found in biologic parents of *another* group of hyperactive children was a necessary compromise. Even with these cautions in mind, the data do suggest a genetic component to the HACS. Further studies of the HACS in the genetic area, such as comparison of identical and fraternal twin pairs and linkage studies, seem to be warranted.

References

Arkonac, O., & Guze, S. B. A family study of hysteria. *New England Journal of Medicine*, 1963, **266**, 239–242.

Cantwell, D. P. Psychiatric illness in the families of hyperactive children. *Archives of General Psychiatry*, 1972, **27**, 414–417.

Chess, S. Diagnosis and treatment of the hyperactive child. *New York State Journal of Medicine*, 1960, **60**, 2379–2385.

Clements, S. D. Minimal brain dysfunction in children. *NINDB Monograph No. 3*. Public Health Service, 1966.

Cloninger, C. R., & Guze, S. B. Psychiatric illness in female criminality: the role of sociopathy and hysteria in the antisocial woman. *American Journal of Psychiatry*, 1970, **127**, 303–311.

Crowe, R. R. The adopted offspring of women criminal offenders: a study of their arrest records. *Archives of General Psychiatry*, 1972, **27**, 600–603.

Feighner, J. P., Robins, E., Guze, S. B., Woodruff, R. A., Winokur, G., & Munoz, R. Diagnostic criteria for use in psychiatric research. *Archives of General Psychiatry*, 1972, **26**, 57–63.

Fish, B. The "one child, one drug" myth of stimulants in hyperkinesis. *Archives of General Psychiatry*, 1971, **25**, 193–203.

Forrest, A. D. The differentiation of hysterical personality from hysterical psychopathy. *British Journal of Medical Psychology*, 1967, **40**, 65–78.

Goodwin, D. W., Schulsinger, F., Hermansen, L., Guze, S. B., & Winokur, G. Alcohol problems in adoptees raised apart from alcoholic biological parents. *Archives of General Psychiatry*, 1973, **28**, 238–243.

Guze, S. B., Goodwin, D. W., & Crane, J. B. A psychiatric study of the wives of convicted felons: an example of assortative mating. *American Journal of Psychiatry*, 1970, **126**, 1773–1776.

Guze, S. B. Tuason, V. B., Gatfield, P. D., Stewart, M. A., & Picken, B. Psychiatric illness and crime, with particular reference to alcoholism: a study of 223 criminals. *Journal of Nervous and Mental Disease*, 1962, **134**, 512–521.

Guze, S. B., Wolfgram, E. D., McKinney, J. K., & Cantwell, D. P. Psychiatric illness in the families of convicted criminals: a study of 519 first degree relatives. *Diseases of the Nervous System*, 1967, **28**, 651–659.

Hoffmann, H. *Der Struwwelpeter: oder lustige Geschichten und drollige Bilder.* Leipzig: Insel Verlag, 1845.

Mendleson, W., Johnson, N., & Stewart, M. A. Hyperactive children as teenagers: a follow-up study. *Journal of Nervous and Mental Disease*, 1971, **153**, 273–279.

Menkes, M. M., Rowe, J. S., & Menkes, J. H. A twenty-five year follow-up study on the hyperkinetic child with minimal brain dysfunction. *Pediatrics*, 1967, **39**, 393–399.

Minde, K., & Webb, G. Studies of the hyperactive child: VI. prenatal and perinatal factors associated with hyperactivity. *Developmental Medicine and Child Neurology*, 1968, **10**, 355–363.

Morrison, J. R., & Stewart, M. A. A family study of the hyperactive child syndrome. *Biological Psychiatry*, 1971, **3**, 189–195.

O'Malley, J. D., & Eisenberg, L. D. The hyperkinetic syndrome. *Seminars in Psychiatry*, 1973, **5**(1), 95–103.

Perley, M. J., & Guze, S. B. Hysteria—the stability and usefulness of clinical criteria. *New England Journal of Medicine*, 1962, **266**, 421–426.

Robins, L. N. *Deviant children grown up.* Baltimore: Williams & Wilkins, 1966.

Rutter, M. *Children of sick parents.* London: Oxford University Press, 1966.

Satterfield, J. H. EEG issues in children with minimal brain dysfunction. *Seminars in Psychiatry*, 1973, **5**(1), 35–46.

Stewart, M. A., Pitts, F. N., Craig, A. G., & Dieruf, W. The hyperactive child syndrome. *American Journal of Orthopsychiatry*, 1966, **36**, 861–867.

Stewart, M. A., Thach, B. T., & Freidin, M. R. Accidental poisoning and the hyperactive child syndrome. *Diseases of the nervous system*, 1970, **31**, 403–407.

Weiss, G., Minde, K., Werry, J. S., Douglas, V., & Nemeth, E. Studies on the hyperactive child: VIII. five-year follow-up. *Archives of General Psychiatry*, 1971, **24**, 409–414.

Werry, J. S. Studies on the hyperactive child: IV. an empirical analysis of the minimal brain dysfunction syndrome. *Archives of General Psychiatry*, 1968, **19**, 9–16.

Woerner, P. I., & Guze, S. B. A family and marital study of hysteria. *British Journal of Psychiatry*, 1968, **114**, 161–168.

VI

SOCIAL IMPLICATIONS OF GENETIC THEORY

PRESIDENTIAL ADDRESS: NATURE AND NURTURE AS POLITICAL ISSUES

Henry Brill

As is well known, the growing importance of the medical and other sciences has led to an increasing penetration of science with political and sociopolitical elements. How to deal with this trend remains one of the major issues of our time.

Some authorities feel that it is desirable and even necessary for scientists to become politically active; others take the opposite view. In either case it seems important to be aware of any possible hazards that may be associated with bringing politics into science or having scientists enter the political arena in their technical capacity. Some will say that only totalitarian politics create hazards and not ordinary politics in a democratic society, and it is true that the clear examples of political damage which we will quote do come from totalitarian states. Nevertheless, there are reasons to examine all such experiences, to study the processes, and to identify whatever danger signals may be found, if they are quite analogous to those which can be observed closer to home when scientists bring political forces to bear on academic and scientific issues.

No scientific area has had a more disastrous experience with this relationship than has the study of genetics. Perhaps because the memory of recent disasters is still fresh, genetics today appears to be in a relatively nonpolitical phase, but the subject still is appropriate for this symposium because the situation is not necessarily permanent even for genetics, and even more because the problem is currently acute in many areas of mental health. I would, accordingly, like to take this occasion to review briefly some details in the story of political genetics and to examine some of the implications which seem to flow from this story. I speak of political genetics, even though the process has been carried out in the name of social advance, because the damage was done by the political forces which were unleashed. I doubt whether we shall be able to draw from our review any strong conclusions as to what the correct path may be, but it certainly shows that genetics and politics can produce a highly dangerous and destructive mixture, and the same may be true of other such mixtures.

Henry Brill, M.D., Director, Pilgrim Psychiatric Center, West Brentwood, Long Island, N.Y.

The political history of genetics appears to go back no more than 175 years. Until that time ideas about nature and nurture had led a peaceful coexistence, and these two factors were recognized as interacting equals in biological processes. Farmers from time immemorial had taken both soil and seed into account in their daily work, and physicians since the days of Hippocrates saw both heredity and environment as important in health and disease (Alstrom 1950). Even the writers of the French Revolution, who were primarily environmentalists, did not develop this as a political issue, and ideas about hereditary disease still remained strong, even if based on the erroneous Hippocratic theory of pangenesis.

The situation changed following the publication of Darwin's *The Origin of Species* in 1859. The full title of the book was *On the Origin of Species by Means of Natural Selection or the Preservation of Favored Races in the Struggle for Life*, a title which in a way foreshadowed the subsequent application of Darwinian theory to social problems.

The Darwinian concepts proved attractive to the intellectuals of the time and quickly gave rise to what came to be known as social Darwinism. Applying the biological concepts of evolution to social processes, competition was seen as necessary to the development of the race; survival was by definition a measure of fitness; and Adam Smith's *laissez faire* laws of economics found reinforcement in biology. Darwin's theories also strengthened certain notions that had long been current about degeneration of the race. These ideas were in part at least derived from the Hippocratic hypothesis of pangeneration, "Semen is excreted from the whole body, healthy from healthy and unhealthy from unhealthy parts." Since mankind was perfect at the start all hereditary defects must have been acquired later and are transmitted in a cumulative way. These ideas were, of course, reinforced by observations in animal breeding which seemed to show that blood lines could indeed degenerate. Morel formalized these concepts in 1857 in a highly influential work (*Treatise on Physical, Intellectual and Moral Degeneration of the Human Species*). For the next half century his ideas and Darwin's dominated psychiatric and social theory. An inherited biological inferiority of the poor was seen as having caused their bad environment and attempts to improve their condition to be futile (Hofstadter 1955); laws and other means of environmental control were counterproductive; progress must be by natural selection and selection must be by competition. The aim of civilization was to maintain competition at all levels in order to achieve progress. Dalton's eugenics became fashionable, and social theory explained the rise and fall of nations by the differential fertility of the fit and the unfit. A long series of studies of the Juke and Kallikak type appeared (Goddard 1939; Deutsch 1937) and seemed to lend further support to the concept that the key to social progress was control of the unfit and not of the environment, which they would create in any case, and that such control, perhaps through sterilization, was the answer to poverty, crime, mental illness, mental defect, and epilepsy (Alstrom 1950).

It should be pointed out again that these ideas, and laws or practices based on such notions, did not appear de novo in the post-Darwinian era. They were deeply rooted in tradition. Sweden's first law forbidding epileptics to marry on eugenic grounds dates from 1757 (Alstrom 1950), and Alström states that it was undoubtedly preceded by less formal restrictions applied through the Church for centuries before. A revision of this law was passed in 1920 and legal restrictions were still in effect at least until 1950. A large number of other eugenic laws directed at sterilization of the mentally retarded and certain other classes were passed in many states of the United States and elsewhere during the early 1900's (Deutsch 1937) and still remain on the books in some localities, although advances in genetic knowledge have made them inoperative for the most part. They are now generally considered to have been premature applications of medical theory through political action.

One can see perhaps some Calvinist remnants (Weber 1958) in these views, with the important difference that superiority or inferiority was attributed to an irrevocable act of nature rather than a manifestation of an irrevocable act of divine providence. These ideas did not long go uncontroverted (Alstrom 1950), and social practice was by no means totally controlled by these theories, but it was impeded.

A far more serious spin-off or derivative of the social Darwinism of the late 1800's was its influence on the Hitlerian philosophy and on Nazi programs when Hitler came to power, and these ideas almost totally controlled national policy. The ideas themselves did relatively little harm as long as they remained largely within the academic and technical context where they had arisen; the damage occurred when they were projected in a total fashion onto a national policy, that is to say when they became truly political. Once scientific theory became a part of political dogma the next phase followed almost inevitably, and opposing points of view in German genetics were suppressed. All that was not pro-Nazi was considered anti-Nazi, and it disappeared from the academic as well as the political scene.

The racist genetics of Nazi Germany was in a sense, counterbalanced by a sort of Russian antigenetics which was to remain dominant from the 1930's until the fall of Krushchev in 1964. Apparently this move was only in part a specific response to the trend in Germany. It was also the result of a broad move to render all sciences relevant to the then current social and political issues in Russia. The story of the way in which the situation developed, as described by Medvedev (1969), has all of the horrifying inevitability of a Greek tragedy. Throughout this account one can identify the danger signals which appeared as science and politics were inter-mingling to produce incredibly bad science and very poor politics. It remains to be seen whether we can look for some of these at home.

It began, says Medvedev, in 1929–31 as a constructive effort at a socialist attack on the scientific front, with the goal of achieving scientific advance. Soon, however, it became a witch hunt for bourgeois tendencies in science, and the teachings of such men as Bechterev in psychiatry and Pavlov in physiology were for the first time labeled as bourgeois, anti-Marxist, and idealistic. The tone of many scientific publications became sharp and sometimes vulgar, and attacks were directed against persons rather than ideas.

In biology a controversy developed between proponents of a "new" theory of inheritance of acquired characteristics and the "establishment" of traditional geneticists. The issue was decided not by scientific data and logic but by determining which view was really consistent with national political dogma, that is Marxism and dialectical materialism. This was the final measure of social and political correctness and relevance in Russia at that time. The atmosphere was one of feverish search for enemies of the people and any scientific discussion tended to become a struggle with political overtones. By 1931–32 many geneticists had come to be considered philosophically as "menshevising idealists" and administra-tive action began to replace them in academic positions with more pro-gressive academicians and to reduce opportunities for research among those who remained.

In 1933–34 the campaign was stepped up after the agronomist Lysenko established a partnership with Prezent, not a geneticist or a scientist but a lawyer turned science teacher and a master of scientific-political demagoguery. Medvedev comments that the human psyche in science runs the full gamut of expression from total mediocrity to absolute genius and includes a range of psychopathic deviations—often more dangerous in the area of talent than of mediocrity. Meetings were now dominated by young activist members who silenced the opposition by the usual tac-tics, and, furthermore, scientists were informed that they could no longer escape by taking a neutral position; there was a class struggle in science, and they would have to take one side or the other. Some of the ideas pro-pounded by the Lysenkoites were intrinsically attractive, as, for example, that there were no hereditary diseases; others were politically attractive, as, for example, the concept that the old genetics was a handmaiden of Goebells, a capitalist genetics that was manhating, arch-reactionary, fascist, and scientific-conservative; it must be replaced by a Marxian genetics of Lysenko. The old genetics was based on the gene which was a nonexistent entity, which had never been seen; hybridization of corn was a useless procedure designed only to serve the interests of the capitalist seed forms; Russian agronomy would alter the nature of plants in a predetermined way by environmental means, because the organism is only a concentrate of environmental conditions, and there are certain periods during which the environment is assimilated into the plant to change its hereditary nature. This neo-Lamarckism incidentally was far from a new idea. It was related to the old pangeneration of Hippocrates and reminds one of the gemmules of Darwin.

In the years which followed, the defeat of the old genetics was completed in all fields of biology, and the details are well worth re-reading, although for our purposes it is enough to identify several characteristics of the process.

1. The major and most dangerous attack came from within the ranks of professionals. It was basically a struggle in which one group allied itself with current political dogma and thus became dominant.
2. No middle or neutral ground was permitted. Compromise was condemned as covert opposition and was punished accordingly.
3. The aim was seen as higher than simple science; it was to seek out and destroy the enemies of the people and all academic opposition was destroyed by turning it into an unpopular political position.
4. It began when scientific controversy became personal and vituperative, and attacks were on men and their motives rather than their ideas alone. It ended with political and social action against the losers, who lost their positions and sometimes their lives.
5. There was an increasing tendency to measure everything by slogans and catch-phrases that had popular appeal. New Robespierres sprang up and dominated the field as self-appointed heralds of social progress and destroyers of the enemies of the people. One is reminded that the old Robespierre's crimes were also committed in the name of social justice. Other high-sounding phrases have been used with terrible effect throughout history; untold millions were destroyed in the name of religion and more millions in the name of improvement of the race. It has been said that patriotism is the last refuge of a scoundrel. But scoundrels are not limited to patriotism, they are flexible and resourceful and any popular phrase or dogma will do. However, it is probably not the scoundrel but the fanatic with a cause who is most to be feared, and then only when he is able to ally himself with political power. As one reexamines the history of those turbulent years in Germany and in Russia one can find many reasons to come to this conclusion which Medvedev reached.

The final upshot of the Russian adventure in political purification of genetics was a series of crop failures and a paralysis of agricultural advance which led to vast losses. Medvedev estimates that in corn alone some 30 to 50 million kilos were lost because of the rejection of American hybridization techniques. In addition Russian genetic science was at a virtual standstill for a quarter of a century.

All of this is now gone, and it would appear that for the present at least genetics is being spared further problems with politics, but it would also seem that on the basis of this experience alone one would be justified in taking a militantly moderate position with respect to the politicising of other branches of public health, including mental health, as well as the medical sciences. The term militantly moderate may seem somewhat of a contradiction, since we usually think of militancy only as applied to extreme positions, but I do not see why moderate positions cannot be maintained with equal force and firmness if not with the same tactics. This is not to deny the need for relevance and responsibility, but it would seem that scientists should recognize their limits with respect to political matters, and live within them. The scientist who moves out of his narrow sphere of expertise has no more weight of authority than does the average citizen; being a poet laureate does not give a man any special qualifications to judge remedies for the common cold. The field of sociology and social psychiatry as it reaches into politics is particularly tempting because it seems that here one can expand his horizons and achieve results far beyond what may be achieved in the laboratory, library, or ward. It appears so much more effective to join a game where the blue chips are on the table and to break out of what seems to be a penny-ante technical operation. The history of genetics, and particularly the Russian debacle with Lysenkoism, shows that this can be a highly dangerous game; the scientist who turns to politics to achieve his aims can do himself, his country, and his profession a vast damage. These principles apply not merely to totalitarian regimes; adaptations of the spoils system and the struggle for power in less extreme form are universal.

References

Alstrom, C. H. *A study of epilepsy in its clinical, social and genetic aspects.* Copenhagen: Ejnar Munksgaard, 1950.

Deutsch, A. *The mentally ill in America.* Garden City, N.Y.: Doubleday & Doran, 1937.

Goddard, H. H. *The Kallikak family. A study in the heredity of feeblemindedness.* New York: McMillan, 1939 (first published in 1912).

Hofstadter, R. *Social Darwinism in American thought.* Boston: Beacon, 1955.

Medvedev, Z. A. *The rise and fall of T. D. Lysenko.* New York: Columbia University Press, 1969.

Weber, M. *The protestant ethic and the spirit of capitalism.* New York: Scribner's, 1958.

21 GENETIC KNOWLEDGE AND HEREDITY COUNSELING: NEW RESPONSIBILITIES FOR PSYCHIATRY

John D. Rainer

There are trends to be noted in psychiatry, as indeed in all branches of medicine. Sometimes these become fads; and in a simplified approach to the history of psychiatric thought one often hears the changing climate described as the swinging of a pendulum. A query addressed to the editor of a medical journal illustrates this point. The doctor wrote as follows:

> At the turn of the century, doctors in our fathers' generation were told that heredity was important in understanding the development of emotional disease and in predicting its likelihood. When I went to medical school, I was told that this was an obsolete theory and that both neurosis and psychoses were due to either childhood experiences, or maybe to neurophysiologic abnormalities. My son is now entering medical school and they are requiring him to study genetics in depth. Is this where my father came in 75 years ago, or is it new and different?

Asked to reply to this letter, I chose to make the following points:

> Both genetics and psychiatry have advanced a great deal in the past decades. Today genetics concerns itself with the interaction between the information contained in DNA molecules ("genes") and the demands or limitations of the environment. From this point of view both health and disease, physical and mental, can be considered as a resultant at any given time of this complex process. To understand the nature of gene expression, the ways in which genes determine the synthesis of enzymes and other proteins, and the regulation of gene activity by environmental factors, medical students today will have to study genetics in depth. There has been sufficient evidence over the past 20 years, from twin studies, family studies and chromosome analysis that much psychiatric illness has a genetic component or predisposition; but unlike 75 years ago, there are new and different ways appearing on the horizon—mathe-

John D. Rainer, M.D., Chief of Psychiatric Research, Department of Medical Genetics, New York State Psychiatric Institute; Professor of Clinical Psychiatry, Columbia University, New York, N.Y.

matical, biochemical, cytological, developmental—for understanding better the interaction of the genetic forces with total life experience.

The fact that in 1973 this venerable and prestigious association has for the first time chosen genetics as the topic for its annual meeting symbolizes the coming of age of this discipline as a basic science in psychiatry. Psychiatric residents throughout the country are seeking and obtaining instruction in the genetic aspects of psychiatry and are learning something of the problems, the methods, the tentative findings, and the direction of heredity research.

In the tradition of one of the small number of pioneering centers in psychiatric genetics in the world, that of the New York State Psychiatric Institute, I would like to give some attention to four current areas of increasing concern. The first of these is the training and education of the psychiatrist in the methods and approaches of genetic research, their usefulness, both actual and potential, and their relation to other disciplines which are part of his education, such as psychodynamics, psychopharmacology, and psychiatric treatment. The second area has to do with the increasing exposure of the general public to both information and misinformation regarding heredity, the search on the part of psychiatric patients and their families for information and guidance in this field, and the corresponding responsibility of the psychiatrist when confronted obliquely or point blank with questions regarding genetic transmission of psychiatric disorders. Third, and allied to this, is the role of psychiatry in exploring the psychological aspects of heredity counseling, and the relationship of the methods and even the personality of the genetic counselor to the effectiveness of his work. In this way, the psychiatrist may contribute to the education of all physicians and other health professionals who are also increasingly confronted with the need to understand, to explain, and counsel. Finally, the profound significance of problems of reproduction and parenthood for the individual and for society, the potentials for good and for harm on the individual and social group level, and the wide range of possibilities for misunderstanding and misuse, create problems of ethics and the rights of individuals which are the topic of much serious discussion and concern today, perhaps second only to the debate over the use of nuclear energy.

Some forty years ago, Ernest Jones (1951a, 1951b) said:

Ever since Mendel's work it has been evident that in estimating the relation of heredity to environment in respect to any character, we have first to ascertain the component units in that character; in other words, what actually constitutes an individual gene. . . . By means of psychoanalysis one is enabled to dissect and isolate mental processes to an extent not previously possible, and this must evidently bring us nearer to the primary elements, to the mental genes in terms of which genetic investigations can alone be carried out. . . . the next study to be applied would be one in the field of heredity [pp. 133 & 160].

Over ten years ago, Sandor Rado (1962) said "We are now standing at the threshold of an era in which the entire proud edifice of medicine— including psychiatry as a whole as well as psychoanalysis—will rest on genetics [p. 259]."

Written in the psychoanalytic tradition, these statements and others by Freud, Heinz Hartmann, Anna Freud, and others should have alerted the psychiatric residents of the postwar generation to the compatibility of the study of inborn and genetic differences with that of the development— genetic in the psychological sense—of the individual within his family. Perhaps it was the deserved fascination, at least in this country, with the burgeoning of interest in the structure and development of the personality and unconscious mechanisms in human motivation; perhaps it was the assumption that the acknowledgment of the role of heredity in predisposition to psychopathology was associated with therapeutic nihilism; perhaps it was the recent memory of the nefarious practices carried out in the name of pseudogenetic theories. Perhaps, indeed, it was also the lack of molecular or mathematical models in which to describe the dynamic interaction between information-carrying molecules and the needs and limitations imposed by the environment, and the lack of any knowledge about the neurological or chemical mediators between the hypothesized genetic forces and the observation of behavior. In any event, only a few decades ago psychiatrists interested in genetics were both rare and lonely.

Today, the psychiatric resident comes to his training with some knowledge of human chromosomes, some knowledge of the genetic mechanism and control of protein synthesis, and perhaps even some knowledge of the distribution and dynamics of genetic polymorphisms in human populations. He is ready to learn and to assimilate to his theoretical frame of reference the data so far presented in his specialty with respect to family and twin studies, investigation of adoptees and their families, longitudinal studies of high-risk children, linkage analysis, pharmocogenetics, and human biochemical and karyotypic differences. He does not have to stifle his curiosity regarding the still unknown intermediate mechanisms of genetic interaction, and he may establish continuity with the pioneer investigators of the past. The overwhelming complexity of psychological and social forces has been acknowledged almost to the point where the search for genetic factors in mental illness may be seen to increase rather than decrease the hope for effective preventive and therapeutic measures.

But the psychiatric resident or the psychiatrist in clinical practice never had and does not now have the luxury of suspending all judgment until the final crucial experiments are done. Patients and their families are also aware of the new findings and the new concepts, though in spite of increasing responsibility and sophistication on the part of science writers they may still develop unwarranted anxiety over premature or inaccurate presentations. It becomes more serious when professional advisors themselves, jumping on the bandwagon, swing as far in the new direction as they may have in the past swung in the old. A woman called me in panic recently saying that a psychotherapist had just told her to have an abortion, since she had a possibly schizophrenic uncle, and her child would therefore have a 50 percent chance of being schizophrenic!

With information now available on the genetic aspects of psychoses and mental deficiency syndromes, proper diagnosis of the condition in question, as well as investigation of family members and interpretation of records and laboratory tests is best assigned to the psychiatric specialist. In the words of John Fraser Roberts (1967), "Genetic advice on mental diseases must be left to psychiatrists. Some of those interviewed, and the histories they give, need psychiatric appraisal. Equally important is the difficulty to anyone not a psychiatrist of interpreting and assessing psychiatric reports [pp. 253–54]." Responsible management of a counseling request requires accurate evaluation of the clinical picture, communication with hospitals and other physicians, physical and mental examinations, psychological investigations, and laboratory procedures. In addition, skill is required to obtain accurate family pedigree material, including information about all family members, even that often repressed or conveniently forgotten, as well as about environmental effects on parents, especially on mothers during pregnancy.

At the present state of knowledge, among the highest risks are those in schizophrenia when both parents are affected, in manic-depressive disorder with probably a largely dominant mode of transmission, in Huntington's chorea with one affected parent, and in some mental deficiency situations such as Down's syndrome in the presence of a known translocation. Even in these situations and certainly in those with a lower though measurable risk, the psychiatrist should be equipped to help the patient and his family evaluate the future personal burden—a quantity which is the result of many factors including the genetic risk, the nature of the clinical condition, the couple's potential as parents, and the effect of a child healthy or unhealthy upon the parents' own difficulties.

By extension these factors have to be considered in genetic counseling in all specialities. A few recent studies have tried to explore the use made of counseling by clients who come to the suddenly proliferating clinics and departments in hospitals and schools throughout the country and abroad. A group at Johns Hopkins (Leonard, Chase, & Childs 1972) recently pointed out that reproductive attitudes of families who received genetic counseling were determined more by the aforementioned sense of burden imparted by the disease than by knowledge of its risk figures. Only about half of the families had a good grasp of the information given, a finding which was related both to their sophistication in biology and to the physician's personal quality and ability to communicate.

There are no more fateful and emotion-laden decisions in a person's life than the one to marry and the one to assume parenthood. Under no circumstances, therefore, ought the proffering of guidance in this area be dispensed as readily by mail as in person. In order to be of constructive help to persons about to make a realistic choice in coping with their particular family problem, the counselor is expected not only to elicit and evaluate the essential facts bearing on a family decision but also to ponder with them judiciously their fears and hopes, defenses and rationalizations, denials and projections. Although genetic counseling sessions need not be categorized as a formal program of psychotherapy, they will greatly gain in effectiveness if based on psychological understanding and conducted along established lines of psychiatric interviewing. In essence, sessions of this kind amount to forms of short-term psychotherapy aimed at bolstering the decision-making apparatus and reducing marital anxiety and tension, shame and guilt, depression, hostility, and sexual inhibition.

A basic difference of opinion, one to which psychiatrists must perforce address themselves, is whether genetic counseling is to be considered merely the provision of information or whether the counselor is to take a role in helping the family reach a decision. Franz Kallmann (1956) pointed out almost twenty years ago that workers in the then limited field of human genetic counseling were inclined to be on the conservative side, the emphasis being placed on tendering a genetic prognosis and no attempt being made to appraise the advisability of parenthood. He made a plea at that time for the need for combining counseling with a personalized effort, saying that in no situation calling for genetic counseling can it ever be taken for granted that the person is realistic or will be able to attain a realistic attitude without the counselor's help. Perhaps, today, with the increased information available, the counselor's attitude may be somewhat less paternalistic, but in my experience clients can no more today than twenty years ago be provided with risk data alone without full consideration of the psychological and social factors which determine the ability to use these data and the conclusions based upon them. At all times the counselor must avoid creating undue anxiety, and he must be well aware of his own emotional biases and reactions so as to avoid interjecting them into the counseling sessions. People who find reason to suspect even remotely the genetic transmission of undesirable characteristics in their families may be more than ever beseiged by guilt, shame, obsessional doubts, and indecision. The psychiatrist has the responsibility for the discussion of these matters in the training of nonpsychiatric personnel as well as their fellow psychiatrists in the psychological aspects of genetic counseling.

Finally, genetic counseling and genetic knowledge carry with them the responsibility for considering ethical and social questions which inevitably arise. Care has to be taken in disseminating information through newspapers and popular journals, so that undue fear is not generated, especially when responsible personal counseling is not immediately available. The same considerations apply to genetic screening programs without adequate facilities for individual guidance.

One important consideration in this area relates to the goal of genetic counseling; is it the immediate good of the family that is being counseled, the health and social well being of the current generations, or the eugenic betterment of the human population? It is not at all clear that the widest goal is feasible, desirable, or necessary. Many geneticists question whether mutations carried in recessive form have only dire results, and emphasize rather that the diversity of genetic potential in the human genotype is what makes the species adaptable to future as yet unknown environments. In their view, the balanced composition of the human gene pool in the long run makes for fitness. Others tend to regard more seriously the load of deleterious mutations being built up in the population, particularly through medical advances. This is not the place to review the arguments and evidence on both sides of this debate; however, the current laudable tendency to place the rights of the individual foremost in medical practice is consistent with the provision of genetic counseling on a voluntary basis and primarily for the betterment of the individual family involved.

Even if screening programs are combined with proper counseling, there are some negative aspects from the point of view of medical and social ethics. Singling out infants with chromosomal defects whose behavioral manifestation is still unclear, such as the XYY karyotype, creates the possibility of establishing a self-fulfilling prophecy, whereby the anxiety of the parents has an adverse psychological effect upon the developing child. In most psychiatric conditions, personal familiarity with the possible vicissitudes and modifications of a genetic predisposition, such as can be gained by the study of discordant twins (Rainer, Abdullah, & Jarvik 1972), is important in providing the balance and perspective for better communication with families coming for help.

In general, as with any advance in knowledge, the coming of age of genetics with its potential for human betterment also requires that there be caretakers who will work to prevent harmful misinterpretation or biological misuse. With increased understanding of how genes and environment interact on individual, family, and population levels, the dissemination of information, the provision and delivery of counseling services, and the integration of such services with other types of health care, marital therapy, and family planning, become essential concerns of a genuinely social psychiatry, and the responsibility of researcher, clinician, and teacher alike. Problems that have to be met in the delivery of genetic services include: invasion of privacy, disclosure of information to other family members, and the status of prenatal diagnosis, of abortion, and of artificial insemination. Other questions that arise have to do with the medical prognosis for an affected child, the difficulties in raising one or more healthy children with the added burden of the sick child, and advice regarding institutionalization, financial aid, and sterilization.

Traditionally, it is not for the psychiatrist to be uneasy in thinking about new and difficult problems. Some of the solutions strike at the root of many human emotions, procreativity, family and social structure. It may very well have to be the job of some psychiatrists to be the caretakers of the human aspects of genetics, to have a voice in the larger issues, and if they do not do the counseling themselves, at least to be concerned, to supervise, and to be involved with it.

References

Jones, E. Mental heredity. (Contribution to a symposium on mental heredity held by the Eugenics Society, January 29, 1930.) In *Essays in applied psychoanalysis.* Vol. 1. London: Hogarth, 1951, pp. 133–134.(a)

Jones, E. Psychoanalysis and biology. (Read before Second International Congress for Sex Research, August 7, 1930.) In *Essays in applied psychoanalysis.* Vol. 1. London: Hogarth, 1951, pp. 135–164.(b)

Kallmann, F. J. Psychiatric aspects of genetic counseling. *American Journal of Human Genetics*, 1956, **8**, 97–101.

Leonard, C. O., Chase, G. A., & Childs, B. Genetic counseling: a consumer's view. *New England Journal of Medicine*, 1972, **287**, 433–439.

Rado, S. (quoted in) Section V, Testimonials and Awards. In F. J. Kallmann (Ed.), *Expanding goals of genetics in psychiatry.* New York: Grune & Stratton, 1962.

Rainer, J. D., Abdullah, S., & Jarvik, L. F. XYY karyotype in a pair of monozygotic twins. *British Journal of Psychiatry*, 1972, **120**, 543–548.

Roberts, J. A. F. *An Introduction to medical genetics.* London: Oxford University Press, 1967.

AUTHOR INDEX

Alanen, Y. O., 174–77

Barchas, Jack D., 27–62
Bender, Lauretta, 125–34, 199–200, 210
Bleuler, Manfred, 175–78, 191
Brill, Henry, vii, 283–88

Cantwell, Dennis P., 273–80
Ciaranello, Roland D., 27–62
Cole, Jonathan O., vii
Crowe, Raymond R., 95–103

Darwin, C., 284
Dunham, H. Warren, 209–15
Dunner, David L., vii

Fieve, Ronald R., vii, 219–32, 241–55
Fink, Max, vii
Fisher, Seymour, vii
Fleiss, Joseph L., 219–32, 241–55
Freedman, Alfred M., vii

Gottesman, Irving I., 167–97, 200, 214

Hamburg, David A., 27–62
Heston, Leonard L., 167–97, 200, 214
Hoch, Paul H., 15, 19
Hutchings, Barry, 105–16

Jacobsen, Bjørn, 147–65, 209–10

Kallmann, Franz, xiii, 210, 219, 230, 235, 241, 293
Kempthorne, O., 137–38, 142
Kessler, Seymour, 27–62, 65–73
Kety, Seymour S., 15–26, 147–65, 183–84, 200–201, 209–10

Kidd, Kenneth K., 135–45, 200, 209–12
Klein, Donald F., vii

Lidz, R., 213
Lindelius, R., 178–79
Lysenko, 286–88

McHugh, Paul R., 233–39
Mednick, Sarnoff A., 105–16
Medvedev, Z., 285–88
Mendlewicz, Julien, 219–32, 241–55
Morel, 284
Motulsky, Arno G., 3–14
Mullaney, J., 259–71

Omenn, Gilbert S., 3–14

Rainer, John D., ix–xii, 241–55, 289–95
Razavi, Lawrence, 75–94
Reich, Theodore, 259–71
Robins, Lee N., 117–22
Rosenthal, David, vii, 147–65, 179–83, 199–208, 209–10

Schulsinger, Fini, 147–65, 209–10
Shagass, Charles, vii
Shields, James, 167–97, 200, 214
Snezhnevsky, A. V., 199
Spitzer, Robert L., vii
Stokes, Peter E., 233–39

Thudichum, 15–16

Von Greiff, H., 233–39

Wender, Paul H., 147–65, 209–10
Winokur, George, vii, 259–71

SUBJECT INDEX

Adoption method, xi, xiv, 95–102, 105–15, 119–22, 147–64, 173–74, 179–84, 201–7, 273–79
Adrenocorticotropic hormone (ACTH), 31–40, 39
Affective disorder, 9–10, 152–53, 276–77
  bipolar, 242–43
  subgroups, xv, 48
  unipolar, 242–43
  See also Depression; Manic depression
Aggression, 75–91, 118. See also Antisocial behavior
Alcoholism, 11–12, 112, 259–71, 276–77
  environmental model, 267–69
  race differences in, 259–71
  two-threshold model, 266–67
American Psychopathological Association, ix, xiii
Amniocentesis, 9
Aneuploidy, sexual, 75–91
Anger, adaptive function of, 41–45. See also Rage
Antisocial behavior, 75–91, 100, 119, 120, 120–22, 128. See also Criminality; Sociopathy
Autism, 128, 129

Background, ethnic-religious, 127. See also Socioeconomic status
Bellevue Hospital, 125–26
Biochemical genetics, 3–12

Catecholamine neuron, 49
Catecholamines, 27–54
Catecholamine synapses, 19
Catecholamine synthesis, 27–54
  genetic differences in mice, 32
Chromosomes, 65–71, 75–91
Color blindness, 241–53
  deutan, 245–46
  protan, 243–44
Co-parent study, 201. See also Adoption method
Counseling, genetic, 71, 89–90, 117–18, 289–95
Criminality, ix–x, 65–71, 76, 81, 90–91, 95–102, 105–15, 117–22, 182. See also Antisocial behavior; Sociopathy
Cytogenetic studies, 65–71, 75–91

Darwinism, social, 284–85
DBH (dopamine-beta-hydroxylase), 21–22, 37–38
DβO (dopamine-β-oxidase), 30
Denmark, 105–6, 115, 121–22, 148–50
Depression, 41–50
  adaptive function of, 41–45
  bipolar, 220–21
  unipolar, 220–21
  See also Affective disorder; Manic depression
Dermatoglyphic studies, 84–91, 119
Diabetes mellitus, analogous to schizophrenia, 164, 192
Diagnostic problems, xi
DNA, 4
Dopamine, 21–22, 30
Dopamine-beta-hydroxylase (DBH), 21–22, 37–38
Dopamine-β-oxidase (DβO), 30
Down's syndrome, 292
Drosophila, gynandromorphic, 88
DSM II (Diagnostic and Statistical Manual of the American Psychiatric Association), 148, 169, 202

Enzymes, 5–6
Epilepsy, 112, 128. See also Seizure disorders
Epileptics, temporal lobe, 78, 81–82, 84–88
Epinephrine, 17
Eugenic laws, 285

Family study method, 125–34, 174–79, 221, 233, 241–53, 259–71
Farnsworth-Munsell-100-Hue test, 243
Fighting behavior, 48
Fingerprints. See Dermatoglyphic studies
Fostering studies. See Adoption method

Genealogy, 222–27, 235–38, 243–46
Gene control of enzyme levels, 37
Gene locus, 135, 192

Hardy-Rand-Ritter test, 243
Hardy-Weinberg ratios, 136
Hitlerian philosophy, 285

299

Homocystinuria, 8
Human genetics, methodology in, xiii–xiv, 201
Huntington's chorea, 172, 292
Hyperactive children, 11, 273–79
Hyperactive child syndrome (HACS), 273–74
Hyperkinetic children. *See* Hyperactive children
Hysteria, 121

Inheritance, mode of
    genetic (single locus), 136–39
    heterogeneous, 230
    multifactorial, 135–45, 259–71
    polygenic, 188–91, 235, 238–39
    selection model, 140–41
    single-gene, 270
    single-major-lucus, 135–45
    two-threshold, 266–67
    X-linked, 36, 88, 192, 250–51
Interviewer bias, 164

Lesch-Nyhan syndrome, 192
Life events, 44–46
Limbic system, 84
Linkage, x, xiv, 219–53
    definition of, 241–42
Lithium Treatment and Research Clinic, 220
Lobectomy, temporal lobe, 88–90
Lod score method, 227–30, 246–53
Lysenko affair, 286–88

Male homosexuality, x
Mania. *See* Manic depression
Manic depression, x, 112, 180–81, 201, 219–53, 292. *See also* Affective disorder; Depression
MAO (monoamine oxidase), 6, 10, 22
Marxian genetics, 285–88
Maudsley Hospital, 186–89
Medical training, genetics in, 289–91
Meiosis, 222
Mendelian principles, xiii
Mental retardation, 128, 129, 179
Methionine, 18
Monoamine oxidase (MAO), 6, 10, 22
Mosaicism, ix, 83, 117
Mosaics, 75, 77, 79, 88, 89
Multifactorial model of inheritance, 135–45, 259–71
    liability in, 260
    See also Polygenic model of schizophrenia
Multifactorial traits, 135

National Institute of Mental Health, 16, 23
Nature-nurture controversy, ix, xiv
NE (norepinephrine), 9, 30
Neuroregulatory agents, 27–54
New York Hospital, 233
New York State Psychiatric Institute, 220, 290
Norepinephrine (NE), 9, 30

Obsessive-compulsive traits, 190
Occupational status, 110. *See also* Socio-economic status

Parkinson's disease, 19–20
Pedigree. *See* Genealogy
Penetrance, 136–37
Phenothiazines, 20
Phenylethanolamine N-methyl transferase (PNMT), 31–41
Phenylketonuria (PKU), 7
PNMT (phenylethanolamine N-methyl transferase), 31–41
Political genetics, 283
Polygenic inheritance of manic depression, 235, 238–39
Polygenic model of schizophrenia, 188–91. *See also* Multifactorial model of inheritance
Prenatal diagnosis, 9
Protein variation, 4–5
Psychopathic, 128
Psychopathy, ix–x, 95–102, 112, 117–22. *See also* Antisocial behavior; Criminality; Sociopathy
Psychotherapy, 293

Rage, 83. *See also* Anger
Recombinants, 222
Recombination fraction, 227, 246–53
Records, arrest, 96, 98–99
    criminal, 106
    psychiatric hospital, 96, 99–100
Reliability, inter-rater, 157
RNA, 4

Schizoid personality, 155, 167–94, 209–10
    definitions of, 168–69
Schizophrenia, ix, 10–11, 20–23, 88, 112, 125–215, 228, 292
    biological substrates of, 15–23
    genetic model, 136–39
    selection model, 140–41
    selective forces maintaining, 135–45
Schizophrenia spectrum, xiv, 125–34, 147–64, 167–94, 199–207, 210

Schizophrenic children, mothers and families of, 130–33
Schizophrenics, 121
Seizure disorders, 11. *See also* Epilepsy
Serotonin, 9, 47
Sex effect, in alcoholism, 264–66
Sex hormone levels, 79
Sexual offenders, 75–78
Sickle-cell hemoglobin, 192
Single-gene inheritance, 270
Single-major-locus (SML) models, 135–45
Social class, 115, 182. *See also* Socioeconomic status
Socioeconomic status, 97, 119–20, 127, 128. *See also* Occupational status; Social class
Sociopathy, 128, 182, 276–77. *See also* Antisocial behavior; Criminality
Species' survival, 42–43
Statistical models, 135–45, 211–12, 227–30, 246–53, 259
Stress, 40–41, 45–46, 51–52, 213

Suicide, 112, 234

TH (tyrosine hydroxylase), 29–37
Transmethylation hypothesis, 17–18
Transmission. *See* Inheritance, mode of
Turner's syndrome, x, 85–86
Twin(s), xiii, xiv, 65, 96, 147, 173, 184–89, 219–20, 243
Tyrosine hydroxylase (TH), 29–37

X-chromosome, 184
Xg blood group, 219–30
X-linkage, 36, 88, 192, 219–53
XXX karyotype, 82
XXY karyotype, 67, 75–84, 89
XYY karyotype, x, 65–71, 75–84, 89, 117–19, 294

Y chromosome, 65–71

Publications of the American Psychopathological Association

Vol. I	(32nd Meeting):	*Trends of mental disease.* Joseph Zubin (Introduction), 1945.*
Vol. II	(34th Meeting):	*Current therapies of personality disorders.* Bernard Glueck (Ed.), 1946.
Vol. III	(36th Meeting):	*Epilepsy.* Paul H. Hoch and Robert P. Knight (Eds.), 1947.
Vol. IV	(37th Meeting):	*Failures in psychiatric treatment.* Paul H. Hoch (Ed.), 1948.
Vol. V	(38th Meeting):	*Psychosexual development in health and disease.* Paul H. Hoch and Joseph Zubin (Eds.), 1949.
Vol. VI	(39th Meeting)	*Anxiety.* Paul H. Hoch and Joseph Zubin (Eds.), 1950.
Vol. VII	(40th Meeting):	*Relation of psychological tests to psychiatry.* Paul H. Hoch and Joseph Zubin (Eds.), 1951.
Vol. VIII	(41st Meeting):	*Current problems in psychiatric diagnosis.* Paul H. Hoch and Joseph Zubin (Eds.), 1953.
Vol. IX	(42nd Meeting):	*Depression.* Paul H. Hoch and Joseph Zubin (Eds.), 1954.
Vol. X	(43rd Meeting):	*Psychiatry and the law.* Paul H. Hoch and Joseph Zubin (Eds.), 1955.
Vol. XI	(44th Meeting):	*Psychopathology of childhood.* Paul H. Hoch and Joseph Zubin (Eds.), 1955.
Vol. XII	(45th Meeting):	*Experimental psychopathology.* Paul H. Hoch and Joseph Zubin (Eds.), 1957.
Vol. XIII	(46th Meeting):	*Psychopathology of communication.* Paul H. Hoch and Joseph Zubin (Eds.), 1958.
Vol. XIV	(47th Meeting):	*Problems of addiction and habituation.* Paul H. Hoch and Joseph Zubin (Eds.), 1958.
Vol. XV	(48th Meeting):	*Current approaches to psychoanalysis.* Paul H. Hoch and Joseph Zubin (Eds.), 1960.
Vol. XVI	(49th Meeting):	*Comparative epidemiology of the mental disorders.* Paul H. Hoch and Joseph Zubin (Eds.), 1961.
Vol. XVII	(50th Meeting):	*Psychopathology of aging.* Paul H. Hoch and Joseph Zubin (Eds.), 1961.

*This volume was published by King's Crown Press (Columbia University). Volumes II through XXVI were published by Grune & Stratton. Volumes XXVII through XXIX were published by The Johns Hopkins University Press.

Vol. XVIII	(51st Meeting):	*The future of psychiatry*. Paul H. Hoch and Joseph Zubin (Eds.), 1962.
Vol. XIX	(52nd Meeting):	*The evaluation of psychiatric treatment*. Paul H. Hoch and Joseph Zubin (Eds.), 1964.
Vol. XX	(53rd Meeting):	*Psychopathology of perception*. Paul H. Hoch and Joseph Zubin (Eds.), 1965.
Vol. XXI	(54th Meeting):	*Psychopathology of schizophrenia*. Paul H. Hoch and Joseph Zubin (Eds.), 1966.
Vol. XXII	(55th Meeting):	*Comparative psychopathology—Animal and human*. Joseph Zubin and Howard F. Hunt (Eds.), 1967.
Vol. XXIII	(56th Meeting):	*Psychopathology of mental development*. Joseph Zubin and George A. Jervis (Eds.), 1968.
Vol. XXIV	(57th Meeting):	*Social psychiatry*. Joseph Zubin and Fritz A. Freyhan (Eds.), 1968.
Vol. XXV	(58th Meeting):	*Neurobiological aspects of psychopathology*. Joseph Zubin and Charles Shagass (Eds.), 1969.
Vol. XXVI	(59th Meeting):	*The psychopathology of adolescence*. Joseph Zubin and Alfred M. Freedman (Eds.), 1970.
Vol. XXVII	(60th Meeting):	*Disorders of mood*. Joseph Zubin and Fritz A. Freyhan (Eds.), 1972.
Vol. XXVIII	(61st Meeting):	*Contemporary sexual behavior: critical issues in the 1970s*. Joseph Zubin and John Money (Eds.), 1973.
Vol. XXIX	(62nd Meeting):	*Psychopathology and psychopharmacology*. Jonathan O. Cole, Alfred M. Freedman, and Arnold J. Friedhoff (Eds.), 1973.

Also published under Association auspices: *Field studies in the mental disorders*. Joseph Zubin (Ed.), 1961.